Land-Grant Universities and Extension | *into the 21st Century*

The image on the cover and on this page is used with permission from the Norman Rockwell Family Trust.

Land-Grant Universities and Extension | *into the 21st Century*

Renegotiating

or Abandoning

George R. McDowell

a Social Contract

Iowa State University Press / Ames

George R. McDowell is a professor in the Department of Agricultural and Applied Economics at Virginia Polytechnic Institute and State University in Blacksburg, Virginia. Prior to becoming an academic, he worked in international development, first in South Vietnam as a volunteer with International Voluntary Services and later as staff for the Peace Corps in Kenya and Malaysia. He has also lived and worked in Zambia and Albania.

© 2001 Iowa State University Press
A Blackwell Science Company
All rights reserved

Iowa State University Press
2121 South State Avenue, Ames, Iowa 50014

Orders: 1-800-862-6657
Office: 1-515-292-0140
Fax: 515-292-3348
Web site: www.isupress.com

Authorization to photocopy items for internal or personal use, or the internal or personal use of specific clients, is granted by Iowa State University Press, provided that the base fee of $.10 per copy is paid directly to the Copyright Clearance Center, 222 Rosewood Drive, Danvers, MA 01923. For those organizations that have been granted a photocopy license by CCC, a separate system of payments has been arranged. The fee code for users of the Transactional Reporting Service is 0-8138-1918-0/2001 $.10.

♾ Printed on acid-free paper in the United States of America

First edition, 2001

Library of Congress Cataloging-in-Publication Data

McDowell, George R.
    Land-grant universities and extension into the 21st century : renegotiating or abandoning a social contract / George R. McDowell.—1st ed.
    p. cm.
    Includes bibliographical references and index.
    ISBN 0-8138-1918-0 (hardcover); 0-8138-1914-8 (paperback)
    1. State universities and colleges—United States. 2. Agricultural colleges—United States. 3. Agricultural education—United States.
I. Title.

LB2329.5.M39 2001
378′.054′0973—dc21
                                       00-063463

The last digit is the print number: 9 8 7 6 5 4 3 2 1

Contents

	Foreword	vii
	Preface	xv
1	Introduction	3
2	The Land-Grant University Interest in Public Service	15
3	The Academy, Science, and Service	28
4	From Theory to Practice in the Agricultural Sciences	48
5	Cooperative Extension—Part of the Problem or Part of the Solution?	65
6	Agricultural Extension—How Well Do the Hostages Serve the Hostage Takers?	83
7	How Cooperative Extension?—The County and Federal Partners	97
8	Promises and Possibilities	131
9	Imagining Extension in an Engaged Land-Grant University	169
10	Conclusion—Renegotiating or Abandoning a Social Contract	191
	References	199
	Index	207

Foreword

By Paul A. Miller[1]

George McDowell's book presents a provocative challenge in a style that is bold, forthright, and polemical. These qualities sparked a disturbing surprise in me on first reading, despite my seasoning as once being a 4-H member, a county agricultural agent, an extension specialist, director of a state Cooperative Extension Service, and president of a land-grant university. But reading further revealed an author whose passion for his subjects grew from experience "on the ground" and underwrites honest and bold conclusions. Disagreements are sure to be aroused; I hope they will prove to be of equal compassion. McDowell minces no words in saying that a predicament resides in the land-grant universities, one that deserves strong language, genuine intention, debate, and reform.

No shortage of commentary about human learning characterizes the present day. Much of it is stirred by the inventions that make it possible to democratize and decentralize knowledge in ways never before dreamed possible, and asks how the university of the future will serve society when it has access to intellectual capital on a world scale. This book also profits from such enunciations as those of the National Association of State Universities and Land-Grant Colleges. As does this author, they also startle the reader when they point out that in the event universities fail to better engage with society, they risk their own obsolescence. McDowell reacts to such indictments in his *Land-Grant Universities and Extension into the 21st Century;* he pounds the table with his concern that revisions in the outreach functions of the land-grant universities have been postponed too long.

Some readers may deny that the author's findings apply to their home institution or even to the land-grant system. Others may shrug that he has generalized too much from too little and is unaware of what is in progress or that crankiness colors his boldness. Whatever the reaction, the author seeks to elicit more penetrating debate by describing the contour of the land-grant system's outreach activities over time, comparing them to unmet needs, and suggesting specific action steps. He strives also to anchor his claims in a larger universe of inquiry.

Academic institutions have survived through one problem-filled epoch after another: reconciling knowledge with religion at the outset; getting science into the curriculum; and adding research as an extraordinary priority. However, the impact of the information era on humankind is likely to exceed the challenges of these previous upheavals. A big question looms: How will the cumulative legacy of traditional universities fare in this new era: the devotion to profession of early Bologna; the independence of scientific research at Berlin; the training of gentlemen and statesmen at Oxford; and the proving ground for technologists at Zurich? In the call for change in those eras, two issues were ever present. One deals with the degree of independence allowed the university by society. The other refers to the scope of what the university offers in its programs. As they have been in the past, these issues are also central in McDowell's treatment of how land-grant universities respond to social need and pressure.

At the 1911 annual meeting of the American Association of Agricultural Colleges and Experiment Stations (the forerunner of today's National Association of State Universities and Land-Grant Colleges), those present were not of one mind on how to invent and install a county agricultural agent as a leader of change in the countryside. On the issue of independence, the association's president exclaimed, "It is seriously to be doubted whether popular conceptions of the aims and methods of education and inquiry are a safe basis on which to establish the policy that shall dominate the work and the influence of either the college or the station." On the matter of program scope, the then Assistant Secretary of the U.S. Department of Agriculture rose to say, "This association should not forget the great importance of other than agricultural lines of endeavor. There are twice as many people in vocations other than agriculture. Why narrow this question to one of agriculture?" The issues of independence and scope will likely continue as long as universities do; they appear on the modern agenda and, 90 years after such previous debates, they remain alive in the land-grant universities and extension. This book serves to define the nature of their present incarnation.

A sense of urgency pervades McDowell's treatise as these issues loom in today's world. This book acclaims the roles of the land-grant universities and extension as two of the most innovative gifts of America to education (a quite plausible claim when admitting ancient China as the initiator of public school education and Europe of the earliest universities). Much can be said of what these universities and extension have contributed to human development and welfare. As the 20th century closes, one may argue that first rank of all its compelling scientific achievements (the revelations and applications of physical laws, the electronic/digital inventions, and dazzling understandings of the atom and the cell) might well go to that system, with the land-grant universities at its center, that served to create and transfer science-base technology into use by agricultural producers. A certain awe attaches to the minuscule fraction of the U.S. population (2 percent presently) that now produces enough food for itself and the rest of the population, exports a fourth of the total to other nations, and has an abundance remaining.

Analysts who deign to explain this miracle confront a complex maze of lay and professional institutions. McDowell, an experienced economist who brings institutional factors into his thinking, describes the statutory and traditional practices in which the land-grant universities and extension are embedded. He also defines how persons in their specific roles interact with each other in order to activate the historic agreements between the counties, states, and the federal government. The author goes beyond describing the general taxonomy of such entities as the state Agricultural Experiment Stations, the Cooperative Extension Services, and related agencies and institutions. He takes the reader into the meticulous interactions of the campus-based research scholar, the extension specialist who links the scholar's research to problems in the field, and the county agent who facilitates its use and adaptation by local people.

Some readers, uninitiated to rural culture, are sure to speculate on the book cover's symbol of the county agent. Were this role better known for its historical ingenuity and importance, the county (agricultural) agent would join the cowboy, the northwoods guide, and the circuit-riding religious pastor as another vibrant symbol of the American experience. The saga of the county agent role reveals how its incumbents helped transform a rural society into an urban one, became models for stimulating improvement in less developed nations, and are now challenged by the likes of McDowell to be part of the cadre to lead in the information era.

George McDowell observes that the service provided by extension—the offspring of the creations of the land-grant universities in 1862, the agricultural experiment stations in 1887, and the Cooperative Extension Service in 1914—reached its pinnacle, a "golden age," in the 1950s, and thereafter began a decline in its ability to accommodate the very changes that it helped bring about. Herein is the author's thesis: In the past three or four decades, the efficacy of the land-grant universities, through extension's efforts to help people solve problems with science-based knowledge, grew less capable to define changes, react to the consequences of earlier work, and address new needs in society. McDowell laments that despite calls for reform, these institutions faltered, resulting in a loss to them and the nation.

He is concerned that the learning opportunities generated by cybernetic advances may overrun and outdistance these institutions as boundaries collapse among all educational institutions, mega-universities appear, and other distance learning venues challenge every college and university. The author does not pretend to predict the future shape of universities. But he believes that the success of the land-grant universities in absorbing and utilizing the electronic future will be measured to considerable degree by how they may update and expand their core philosophy, once if not now embodied in extension, to the entire institution.

Land-Grant Universities and Extension into the 21st Century employs several modes to support the author's analysis, including a concise history of the land-grant universities and the Cooperative Extension Service, definitions of public service, its several forms, and the benefits to provider and client. Noting the tensions between the university, science, and the public, and their effect on the major players helps the author to identify features of the academy that may facilitate or hinder its practice. To demonstrate these analyses, the author turns to the case study, including close scrutiny of his own discipline of agricultural economics and of the university that he serves.

Following the narrative and case studies, McDowell imagines how a land-grant university might appear if created anew in light of today's world. He rejects tinkering in the absence of imagination to avoid the illusion of reform. By starting fresh, McDowell can lay his big cards on the table with strong language. He has no broad empirical study of the land-grant system at hand, and some may contest that his cases are too limited and parochial. But his combination of history with specific cases indicative of more general conditions join with statistical summaries to support his conclusions. Anchored in his experience and scholarship, his imagination goes to work and forges a courageous statement,which asserts that extension got stuck by playing safe, drifting between a diminishing rural society and the urban transformation, while the parent universities stood by with too little interest and leadership!

Given his subject, McDowell is not confined to only the outreach function of the land-grant university. He probes its interior regions as well. Of special interest is his review of the "public commission" of the university. Drawing on one of his mentors, philosopher John F.A. Taylor, and leading into the epistemological aspects of knowledge discovery and application, he explores how faculty are trained, located, tenured, and rewarded for the outreach function; the strengths and weaknesses of missions attuned to technology transfer; and the necessities and constraints of partnerships with governmental groups and other institutions and agencies. These concepts and references make the book an insightful guide for exploring the linkages of public universities to society.

Returning to the pivotal issues of macro-changes in the basic nature of universities—independence and scope—certain of the author's points underwrite large reforms. He recognizes, however, that a thin line separates changes, which may go too far and too soon or too little and too slowly. This delicate adaptation of the university's institutional independence and scope of program competence to societal need are of major interest. As he focuses on the land-grant universities and extension, three major concerns join to shape his analysis.

First, McDowell asserts that extension is "held hostage" by two historic sponsors and helpmates—the general farm organizations and, also in a symbolic way, the U.S. Department of Agriculture (USDA). This metaphor illuminates the failure of extension to adapt its mission, resources, and practice to major consequences of its earlier and remarkable contributions—such results as the commercialization of American agriculture and the urbanization of society. He also notes the general reluctance of the land-grant university system, with some exceptions, to comprehend this lethargic adaptation and join extension in revising the latter's mission, organization, and program. In addressing this problem, McDowell follows others. As noted earlier, as far back as 1911, those who took the long view were concerned that the real and symbolic relationships of extension and the universities to the farm organizations and the USDA would eventually limit sponsorship, timely advisory modes, and program relevance.

McDowell reminds us over and over of the success of the land-grant university in its use of institutional freedom in the course of a long history. But in recent years, and at a quickening pace, he describes how these traditional orientations have held so firm as to limit responses to new challenges for public service: marketing and other issues in the food industry, the weakening of rural institutions, and the cry for help in facing urban problems. Noteworthy steps to break out of these constraints were taken, especially with orientations to management, resources planning, and community services and institutions. However, in McDowell's view, these initiations on the whole were, if not token, sufficiently inadequate to leave extension and its research base adrift in the wake of the American transformation. Even McDowell's own discipline of agricultural economics does not escape a hard look at its limited place and contribution.

Second, McDowell moves from limitations imposed by advisors and sponsors into the consequences for program scope and related outreach efforts in need of reform. He cites factors that bear on the meaning of scholarship in the future and leads the reader into considering the interplay of science and service and of theory and practice. He deals with content of program that the world outside the academy needs, then turns to how the university's disciplines not only fail in their responses but grow isolated from both the problems searching for solutions and from each other.

This work adds to a long history of concern with what extension programs should include, already documented in a litany of reports, conferences, and technical papers urging broadened scope; at least one classic document emerges per decade. But despite these official intentions, McDowell reminds us that when taking the whole system into account, by summing up personnel and budget allocations, and despite laudable exceptions in specific situations, a broader scope of effort appears more fictitious than factual: technology transfer to production agriculture still dominates the mix!

The reluctance to modify extension's role so that the entire university might join in the outreach function also remains a disappointment. This conclusion partly explains why extension's future seems so threatened. McDowell is quick to express his own concern over this limitation and its consequences and takes the analysis into even deeper regions. Among the recommendations is his urging of universities to adopt the broader meaning of scholarship exemplified in Ernest Boyer's *Scholarship Reconsidered* (1990), surely one of the most important documents in higher education of the past half century. Boyer spoke of the discovery of knowledge and the need to add other forms of scholarship that focus respectively on its integration, application, and teaching. McDowell uses case examples to indicate the idea is feasible and under way. Alas, however, unless the university world and graduate education join to widen the meaning of scholarship, another disappointment looms for creating the engaged university.

A third sentiment gives overall strength to the book, one that is surprising from an agricultural economist cultivated in the pragmatic ways of farm production, marketing, and policy. But aware of the history of the university as an idea, McDowell accepts the requirement that the university must be both independent of and engaged with society.

Perhaps seeking his own position on this conundrum, he notes how other views on engagement may vary. One is general and conservative, a belief that in all the university does a service is provided, one to be measured in the long term and aimed at the intellectual, cultural, and material uplift of society. Derek Bok's *Universities and the Future of America* (1998) is cited as one such viewpoint.

On the other hand, McDowell turns to the Kellogg Commission's call for more and better outreach services of public universities, as embodied in its several reports, notably *Returning to Our Roots: The Engaged Institution*. He is responsive to the pragmatism and its call for early reform; one may suppose it helped prompt his book. In the end, however, as he imagines how a land-grant university would appear if created today, one senses (as when he warmly joins the social sciences, the arts, and humanities to the natural sciences), that he locates his views somewhere between those of Bok and the commission. Thus one gathers that the engaged university is not meant as an agency, nor should it unwittingly become one where clients' short-term wants, if not needs, rule supreme. With access to learning resources now multiplying, and given all the needs people have, those most easily met are those of vocational skills and similar utilities. A true university, while seeking its share of that market to be sure, must pursue the agenda for which it is uniquely fitted.

George McDowell sees the university (whatever its auspice) being true to itself. He listened well to Taylor (also a mentor to me), whose writing in a single sentence could carry a literary power never to be forgotten, e.g., "The civilized and civilizing risk which the society assumes in creating the university is that it is creating its own critic." McDowell argues that in whatever the university attempts, the well-being of society must be foremost. But service is not to be confused with the rendering of services. While one may begin with services, the basic quest of the university, this book exclaims, is to end with a society in which all have opportunity to learn broadly throughout a lifetime.

The impact on the university of a knowledge-centered era, which all people have opportunity to enjoy, will exceed other periods in the past. As the barriers between learners and knowledge weaken and fall, including those between campus and community, the university of the future is as yet unknown. In this energetic volume, George McDowell shares his own imagination on that future, yet respects and compares other views and examples. He directs this honest and imaginative book to the land-grant institutions and to that "institution within an institution," the Cooperative Extension Service. His is a subject and a style that will be read and will endure.

While this work will upset some and challenge all, McDowell believes that only strong and straight talk can suggest reforms in a university, which, by its fundamental nature and importance to its mission, is cautious in changing. He has written a bold and controversial book, but it does not omit that he appreciates the resilience of the university over long periods of time, and accepts that it not forsake those immutable truths that steady it. This balancing of short-term with long-term responses to society reflects the insight of Taylor and other champions of the university who were similarly aware. Important among them was Ortega y Gassett, who closed his classic *Mission of the University* (1944) by stating, "The university must be open to the whole reality of its time. It must be in the midst of real life, and saturated with it [and] . . . must intervene, as the university, in current affairs, treating the great themes of the day from its own point of view: cultural, professional, and scientific."

Note

1. Dr. Miller was an extension agent in West Virginia from 1939–1942; professor and extension specialist in sociology at Michigan State University, 1947–1955; Deputy Director and Director of Extension, Michigan State University, 1955–1961; Provost, Michigan State University, 1959–1961; President, West Virginia University 1962–1966; Asst. Secretary for Education, HEW, 1966–1968; President, Rochester (NY) Institute of Technology, 1969–1979; among other positions.

Preface

My first encounter with extension was between high school and college in 1957. I was working on the McGuire dairy farm in Washington County, New York, and attended a special winter extension meeting for area farmers on the management of their soils. Dr. Reashon Feurer, Extension Soils Specialist from Cornell, was one of the speakers. I remember Dick McGuire asking about a particular soil type that was being described; he thought that soils in a particular field on his farm were an exception and said so. In response to the question by McGuire, Dr. Feurer, who only later revealed that he had done much of the mapping of the soils in the county, asked to which field McGuire was referring. Dick replied that it was unlikely Dr. Feurer would know it. It was the one he called the lime-kiln field that was at the top of the ridge above Hedges Lake. Dr. Feurer said to the effect—"oh yes, that's the little field you get to through the break in the hedge row off that old road that goes up that ridge—I can see where you would think it is an exception to this classification, but that field is actually a different soil type all together." Clearly the Extension Soils Specialist knew the McGuire farm better than Dick McGuire. And Dick McGuire was no slouch—he subsequently served New York state as Commissioner of Agriculture for almost seven years. I was impressed at the level of knowledge evidenced in that extension meeting.

When in 1975 I became an assistant professor in the Department of Food and Resource Economics at the University of Massachusetts with a 75 percent extension assignment, a 15 percent teaching assignment, and a 10 percent research assignment, I was sure there would be enthusiastic support for my commitment to extension work. I have degrees from three land-grant institutions, the University of Rhode Island, Cornell University, and Michigan State University, and thought I understood a bit about their threefold mission. I was hired to work on rural development and wanted to know those issues in Massachusetts as well as

Reashon Feurer knew New York soils. But the Department of Food and Resource Economics at the University of Massachusetts was in an uproar.

A recently arrived group of young Turks wanted to make the department the best resource economics department east of the Mississippi, the Hudson, or the Connecticut River. One of the group members even proposed that since everyone knew what the best journals were, we could rank them and then annual raises and evaluations would be simple and quantitatively based on the number of pages of articles in the respective journals multiplied by the reciprocal of the journal rank. That measurement had little to do with the knowledge I needed if I were to even come close to Reashon Feurer. Much of what I needed to know would never fit those journals.

A clear and unambiguous message also was being sent to me from the young Turks: Most of the folks who engaged in extension were lazy, and certainly were not "rigorous." While I could do extension if I wanted to, I would have to make the cut on the same basis as they did. Above all, I wasn't to get in their way in the takeover of the department on the way to fame east of whatever river they chose. None of that road to fame had any place for extension on it, at least as they saw it. Acceptance of my work as "different but also valid" was threatening because it suggested that someone might propose that they too should do extension. Fortunately, the new Associate Director of Cooperative Extension, the functional leader in Massachusetts, Gene McMurtry, affirmed my work and my instincts about what extension was about.

After too many fights that were becoming personal, I decided in the early 1980s that I was fighting on the wrong battleground—I would never win in the department. I then started writing about extension and land-grant universities in the *American Journal of Agricultural Economics*, in *Choices*, and for professional meetings. Some of that writing is the basis of several of this book's chapters. It also explains the vintage of some of the references, which were the contemporary literature of the day, and are, as this book was researched, still relevant.

The book was to have been written in 1993 when a representative of Iowa State University Press approached me about doing a book based on my past writing and presentations. At the time, I was fully engaged in preparing for an extended stay in Albania and suggested that I would work on the book during the long evenings after a casual day of work at the Agricultural University of Tirana. No such days or evenings presented themselves and so the book had to await my return from Albania, a new relationship with new staff at Iowa State University Press, and a grant from the Kellogg Foundation through Dr. Gail Imig.

Several people fit the category of "without whose help this book would not have been written." Gail Imig is one of those, as is Rachel Tompkins, who introduced me to Gail and encouraged me. Others in the category are Jim Hite, Sandra Batie, Paul Miller, James Bonnen, Peter Bloome, Clark Jones, and Gene McMurtry. Gene, who died in 1981, was my extension mentor and helped me understand the overall extension system. Darcy Meeker, my editor, was kind and gracious as she helped me tell the story with greater clarity and less opportunity for misunderstanding.

Dozens of others in land-grant universities and other organizations around the country have helped enormously with the insights and details of the land-grant university extension system as described. Thus many citations in the book make reference to a "personal conversation" or "personal communication." Many of those people are included in the list below, but surely I have left out several. To those left out, I apologize for the oversight.

I also acknowledge the contribution of my colleagues in the Department of Agricultural and Applied Economics at Virginia Tech, particularly those involved in outreach. They tolerated my noninvolvement in other departmental activities during the writing of the book, gave me ideas, and acted as sounding boards.

The following 132 people from around the country deserve special mention: Mary Ahearn, David Alexander, Dave Allee, Gene Allen, Ted Alter, Ann Argetsinger, Walt Armbruster, Henry Bahn, David Barrett, Al Beaver, Claude Bennett, Linda Benning, Steve Blank, Dale Blyth, Barbara Board, Paul Bonaparte-Krogh, John Bottum, Charles Boyer, Joe Broder, Leslie Burns, Jeanne Bush, John Byrne, Billy Caldwell, Gerald Campbell, Emery Castle, Daryl Chubin, James Clark, Cindy Clark-Ericksen, Patrick Corcoran, Sam Cordes, Tom Covey, Ellis Cowling, Larnie Cross, Martin Culik, Barbara Cummings, James Dickerson, John Dooley, Mike Dunkin, A.J. Dye, Del Dyer, Merrill Ewert, Katherine Fennelly, Roger Fletcher, Rodney Foil, Tom Geiger, John Gerber, Gordon Groover, Lynn Harvey, Ed Harwood, Mary Heltsley, Bart Hewitt, Jim Hildreth, Beth Honadle, Lyla Houglum, Greg Hutchins, Glenn Johnson, Stan Johnson, Tom Johnson, Myron Johnsrud, Bernard Jones, Eluned Jones, Judith Jones, David Kenyon, Robert Koopman, Mike Lambur, Mark Lederer, Max Lennon, Larry Libby, Richard Liles, James Littlefield, Donna MacNeir, Deborah Maddy, Peter Magrath, Karen Mundy, Trish Manfredi, Peggy Meszaros, Bob Milligan, Beth Moore, George Morse, Les Myers, Angela Neilan, Terry Nipp, Cindy Noble, Richard Nunnally, Carl O'Conner, Jeff Olsen, Phyllis Onstad, John Ort, Jim Pease, Fariba Pendleton, Scott Peters, Sue Pleskac, Ron Powers, Everette Prosise,

Glenn Pulver, Wayne Purcell, Richard Rankin, Barbara Reeves, John Richardson, Tom Riese, David Riley, Rustum Roy, James Ryan, Dick Sauer, Neill Schaller, Lyle Schertz, John Schnittker, Jim Scott, Norm Scott, Ron Shaffer, Stephen Small, John Smart, Ann Sorensen, Judith Stallmann, Garry Stephenson, Barbara Stowe, Paul Sunderland, Gene Swackhamer, Andy Swiger, Ross Talbot, Tami Torquato, Minnell Tralle, Ron Turner, Luther Tweeten, Beth Van Horn, James Votruba, Henry Wadsworth, Bud Weiser, Cleve Willis, Dot Wnorowski, Bill Wood, Jet Yee, and Russ Youmans.

I am especially indebted to the W.K. Kellogg Foundation. Their support made it possible to travel to many land-grant universities, and to attend several conferences relevant to the question of the future of universities and extension.

Finally, the implied analysis of the book comes out of my training in analytical institutional economics and the tutelage of A. Allen Schmid. I have tried to exorcise much of the economic jargon from the writing, but some remains I am sure. As with all analysis and particularly with institutional analysis, where you stand depends upon where you sit. Thus there is much room for individuals sitting in other seats to have different stands than those taken on most of the issues discussed by the book. The test of validity of the analysis depends on whether the arguments, the observations, and the analytical insight ring true to readers who also know the system and can, at least for the moment of the reading, occupy the seating and the standing espoused. Because of that test of the efficacy of the arguments, I have tried to be as accurate as possible about examples used to illustrate points. If errors remain, and thus erroneous conclusions, they are my responsibility alone.

I dedicate this book to my wife, my best friend, Delores. You're the best!

Blacksburg, Virginia

Land-Grant Universities and Extension | *into the 21st Century*

1

Introduction

The American land-grant universities at the beginning of the 21st century include some of the finest institutions of higher education in the world. They also represented a uniquely American approach to democracy—providing for the "vulgarization" of higher education to the "industrial classes" of the society. The Morrill Land-Grant Act of 1862 was, says J.F.A. Taylor, "the charter of America's quietest revolution" (Taylor 1981, 37). The 17.43 million acres of land in the public domain committed to finance the land-grant colleges—30,000 acres per senator and congressman in each state—are not the things to be attended in reflecting on the establishment of these institutions. Rather, it was the principle behind their establishment that was without historical precedent. That principle asserted that no part of human life and labor is beneath the notice of the university or without its proper dignity. Both by virtue of their scholarly aims and whom they would serve, the land-grant universities were established as people's universities. This was their social contract.

As the United States enters the 21st century there are 51 land-grant institutions that were established under the Morrill Land-Grant Act of 1862—one in each of the 50 states and one in Puerto Rico. There are an additional 17 land-grant institutions established or supported, as in the case of Tuskegee University, under a second Morrill Land-Grant Act of 1890. These are the "traditionally black" institutions that suffered considerably under the "separate and unequal" philosophy of education that dominated education for African Americans through much of the 20th century. From 1890 to 1994 there have been an additional 34 institutions created, most of them community colleges for Native Americans and four-year institutions in the U.S. territories including Guam and the Virgin Islands. The University of the District of Columbia was one of the more recent land-grant institutions created.

Extension services in the 50 states and Puerto Rico that have federal support embrace all of the land-grant institutions that are in the state, with joint leadership by the 1862 and 1890 institutions, though the 1862 institutions are dominant in the extension systems.

Prior to the 1862 land-grant institutions, higher education was reserved for, and helped preserve, the aristocracy of the society.[1] Being a university graduate was an imprimatur of high status in the society. The land-grant universities opened classrooms to young people whose previous experiences were primarily in the cow barn, the kitchen, the forge, or the coke oven. Liberty Hyde Bailey, America's preeminent horticulturist, father of the discipline of horticulture in America, and dean of the New York State College of Agriculture at Cornell from 1903–1913, wrote that:

> Education was once exclusive: it is now in spirit inclusive. The agencies that have brought about this change of attitude are those associated with so-called industrial education, growing chiefly out of the forces set in motion by the Land Grant Act of 1862. This Land Grant is the Magna Charta of education: from it in this country we shall date our liberties (Peters 1998, 53).

The idea that academic institutions should reach out to serve the workaday needs of society was not, however, the major motive of Justin Morrill in sponsoring the Land-Grant Act in 1862. It appears, rather, that he was primarily interested in the nexus between democratic access to higher education and the maintenance of political democracy:

> The land-grant colleges were founded on the idea that a higher and broader education should be placed in every state within reach of those whose destiny assigns them to, or who may have the courage to choose industrial vocations where the wealth of nations is produced. . . . It would be a mistake to suppose it was intended that every student should become either a farmer or a mechanic when the design comprehended not only instruction for those who may hold a plow or follow a trade, but such instruction as any person might need . . . and without the exclusion of those who might prefer to adhere to the classics (Morrill 1887).

Indeed, as America enters the 21st century, the national and individual ethic with respect to formal education is dominated by the expectation of access to higher education for all—it has become commonplace. Today we expect all young Americans, who are able, to go to college. Many of them expect to go on to graduate school at least for a master's degree. Even though other developed nations have emulated the U.S. experience in recent years with respect to this investment in higher education, still, during the period from 1985–1991, the United States consistently reported the highest enrollment for 18- to 21-year-olds in tertiary education of all OECD countries[2] with enrollment rates between 33 and 38 percent (Peri et al. 1997).

Thus, in the mid-1990s, according to the Carnegie Foundation, there were 3,595 institutions of higher education in the United States. Fully 42 percent of these institutions were associate of arts institutions, such as publicly funded community colleges, which offer two-year programs and enroll 42 percent of the 15.3 million postsecondary students. With respect to access of ordinary people to classrooms for postsecondary education, the community colleges are replacing the land-grant colleges and universities as the people's colleges. However, the land-grant universities continue to be an important source of baccalaureate and postbaccalaureate education. According to the 1998–99 Almanac of the *Chronicle of Higher Education*, of the 35 universities awarding the most earned doctorates in 1996, 18 were land-grant universities (Chronicle 1998).

However, the "vulgarization of higher education" conveys only a part of the significance of these universities in the history of higher education in the world. There was an even more profoundly revolutionary idea embedded in the establishment and evolution of the land-grant universities than widespread higher education for ordinary citizens. It was, in Taylor's terms, "that thought and action were indivorcible, that the place of the academy is in the world not beyond it, that it is the business of the university to demonstrate the connection of knowledge, art, and practice" (Taylor 1981, 37). Prior to the land-grant universities, the aristocrats of the world and of America were schooled in theology, the letters, law, and, in a few institutions patterned after the German universities like Johns Hopkins University, medicine. The land-grant view of scholarship directly challenged the prevailing norms of higher education at the time of their inception by making the work of cow barns, kitchens, coke ovens, and forges the subject matter of their scholarship. In 1890, the Babcock test for butterfat content of milk was both a scientific advancement and a political/economic act necessary to rationalize markets for fluid milk.

The expanding demand for higher education, including graduate education, but particularly the demand for the knowledge from the applied/empirical brand of scholarship introduced by the land-grant institutions, has led to the growth in both the numbers of institutions and their sizes and scopes since 1862. Of the 3,595 higher education institutions in the United States at the end of the 20th century, the Carnegie Foundation classifies 125 institutions, both public and private, as "research universities." The 125 research universities are the jewels in the crown of American higher education[3] and of those, 43, or fully one-third, are land-grant universities. Of the research universities, 69 are "Research I" institutions and 22 of those are land-grant institutions. The 1862 land-grant universities not classified as "research universities" are grouped within the next classification of "doctoral" institutions.[4]

In commenting on a revision of the Carnegie Classification, Ernest L. Boyer, past president of the foundation stated:

> The overall number of institutions in the 1994 Carnegie Classification increased from 3,389 (in 1970) to 3,595. The new Carnegie Classification also reveals what some have called the "upward drift" in higher education, and of special interest is the continuing expansion of research and doctoral institutions. America must continue to support a core of world-class research centers; they are essential to the advancement of knowledge and to human achievement. Such activity is costly, however, and it is crucial that we have available the fiscal resources needed to sustain an expanding network of institutions devoted to scholarly research (Boyer 1997).

At the end of the 20th century, with most of the land-grant institutions falling into the research university classification and the rest listed as doctoral institutions, it is clear that the dominant influence on their evolution has been their role in generating new knowledge and graduate education, despite very large undergraduate enrollments. Land-grant universities have relinquished some of their early roles of increasing access to formal higher education, and thus have relinquished that part of their social contract to other institutions, such as the community colleges. Maintaining that part of the social contract as their primary relationship with the American people would have perhaps been an impossible task given that in 1994, there were 15.3 million students involved in tertiary education in the United States (Boyer 1997). The land-grant universities have played a crucial role, however, in facilitating that access by training many of the Ph.D. and master's degree faculty for the teaching institutions now providing the major access to higher education in America.

Access to classroom instruction is not, and has not been, the only way in which the land-grant universities fulfilled their contracts with Americans regarding public access to the knowledge they create, though that was the initial effort. After agricultural scientists demonstrated their abilities to solve some of the practical agricultural problems, both the scholarly agenda and the access to knowledge were inextricably entwined at the land-grant colleges around 1900. By this time, farmers, hungry for solutions to their problems, clamored for the insights of the scientists. The claims on scientists' time became so great that the outreach function of the university was formalized as the Cooperative Extension Service by the Smith-Lever Act of 1914.

Smith-Lever provided for federal government funding to the universities in support of the extension outreach function, just as the Hatch Act of 1887 funded agricultural research. Indeed, as Rainsford's (1972) research makes clear, the Smith-Lever Act was passed because the direct benefits sought by agricultural interests in their support of both the Morrill Act of 1862 and the Hatch Act had not been forthcoming. Most students in the land-grant colleges did not study agriculture, even though they came from farm families; results of research and instruction did not reach farmers because they were not in college but on the farm.

The public service mandate of the land-grant universities stemming from the Smith-Lever Act of 1914, rewritten in 1953, remains ". . . to aid diffusion among the people of the United States useful and practical information . . . and to encourage the application of the same (Hildreth and Armbruster 1981)." The Smith-Lever Act is widely interpreted within the colleges of agriculture in land-grant universities to encompass a broad array of subjects that pertain to the problems of individuals, households, businesses, and governments. Most importantly, this earliest mandated public service function in American higher education is an active, usually nonformal, functional education activity based on the scholarship of the university and directed to widely dispersed and varied audiences beyond the campus. The cooperative extension service programs of the land-grant institutions still function in this manner in a wide variety of disciplines.

This institutionalized form of public service has had a profound impact on the character of higher education in America. In describing the importance of this influence on American higher education, Stephen R. Graubard, editor of *Daedalus,* states in the 1997 preface to a *Daedalus* edition devoted to the American academic profession:

> Without wishing to deny the importance of (the influences of the German and British universities), the uniqueness of the American system needs to be emphasized, and not only because of the Morrill Act and the innovations introduced by the land-grant principle, with its emphasis on research in agriculture and many other fields as well. The concept of "service" took on a wholly new meaning in state universities that pledged to assist their citizens in ways that had never previously been considered (Graubard 1997).

The land-grants were to be people's universities. With the extension function in place by the passage of the Smith-Lever Act, the institutionalized access of ordinary people in the states to their state university was provided for with federal leadership. The federal government partner to the system was, and is, the United States Department of Agriculture (USDA) whose approximately $1 billion of funding contributes to research and extension for a predominantly agricultural clientele.

This system that integrated research and extension has been, and is, hugely successful—for agricultural productivity, for the farmers who have survived the economic tests, and for American society—though many farmers have been made obsolete by the system's results. The rate of return on investments in research and development and extension in agriculture is somewhere between 20 and 40 percent per annum. In a society whose long-term cost of government borrowing has seldom if ever been as high as 15 percent, it could be argued that government should borrow at 15 percent and gain returns of 20 percent by investing in agricultural research and extension (Alston and Pardey 1996).

Evidence of the success of the system was clear by the period from 1920 to the end of World War II, called the "Transition to Science" era of American agriculture by Huffman and Evenson. It was during this period that hybrid corn, among other science-based advances, was developed. However, the period of the 1950s and 1960s was the Golden Age for the land-grant agricultural research and extension system according to Huffman and Evenson. By that time, the system was enabling U.S. farmers and the agricultural sector to successfully compete with producers anywhere in the world, as well as be judged as one of the most productive sectors of the U.S. economy (Huffman and Evenson 1993).

There are reasons to believe that the engagement of campus-based scholarship with the realities of agricultural problems at the farm level through extension is part of the explanation of the huge productivity of the system. Notwithstanding the low emphasis given to the extension function by writers on the system's economics (Huffman and Evenson 1993, and Alston and Pardey 1996), the extension function is certainly a necessary if not sufficient condition to the system's success. There was a time during this period that extension was adjudged to be the most trusted source of new knowledge for ordinary Americans—clearly an affirmation of the efficacy of the system.

In 1999, there were more than 15,000 full-time equivalent (FTE) extension staff associated with land-grant universities with offices in virtually every county of the country—actually there are not county offices in Massachusetts and Connecticut, but they may be the only exceptions. Many extension staff have university faculty status whether they are located on the campus of the state's land-grant university or in a county office. In the colleges of agriculture, which have the longest tradition with the formal extension function, large numbers of faculty members in academic departments have responsibilities for extension as well as teaching and/or research activities. In 1997, estimated total expenditures for this outreach function of the land-grant universities from all sources was about $1.5 billion of which federal funds were about 25 percent (USDA 1997).

But support for extension from both state and federal levels has generally been under assault at the end of the 20th century, and in many states there has been some decline and even greater threat of loss of support for extension in the past 25 years. The reason is quite simple and clear. The success of the agricultural research/extension establishment and the increased productive capacity of farmers made it possible to produce the nation's food with ever fewer farmers. Indeed, as more successful farmers survive and less successful farmers go out of business, farm business size has grown and farm numbers have declined (USDA 1999). In 1997 with 2.05 million farms, there were almost 400,000 fewer farms (16 percent decline) than in 1977. But the extension portfolio of programs has not followed suit. Indeed, during the same period, the proportion of extension resources committed to agricultural programs has grown rather than shifted toward new clients and new problems.

The strong support by farming people for the land-grant system via extension, particularly during the Golden Era, was support for educational programs directed to their farming needs. They were not supportive of the system for its own sake or for its general value to the society. Indeed, farm groups in many states became quite possessive of colleges of agriculture and of cooperative extension. With declining political influence on the part of farm groups, the size of the extension budget was not maintained at the levels attained during the 1950s and 1960s. However, farm interest groups still exerted sufficient power in the system to attempt to assure that their programs would be continued without disruption despite budget declines. Thus, just when the extension system most needed to shift resources to serve other clients, thereby developing additional constituents and support to grow the total budget, agricultural interests' demands assured an ever-declining budget by their efforts to influence or control the internal allocations of the system.

The land-grant universities, particularly those that are research universities, have grown to become enormously diverse and complex institutions, vastly different than the "Land-Grant Colleges of Agriculture" they once were. While colleges of agriculture or their successors by whatever name they are known still play an important role, there is much more than agriculture in the university that is of interest and impact to the people of the state that support it. Notwithstanding the changes in the university and the profound changes in the society, the portfolio of extension at most of the land-grant universities at the end of the century more closely reflects the problems of the society during the Golden Age of agricultural research and extension in the middle of the 20th century than it does the society at the end of the century.

America, as she enters the 21st century, is a high-technology society. Because the research universities are a major source of new knowledge in that society, and because there is an American ethic of "can do," the new knowledge of the research universities will find its way into the hands of the American people, with or without programs to facilitate the flow of that knowledge. However, there is an organization with offices in virtually every county in the country with formal affiliation and funding from the land-grant university of the respective states. The extension organization has performed in a stellar manner in the past. It seems obvious on the face of things, that it should play a role in facilitating the flow of knowledge from these crown jewels of the public higher education system to the people of the society. But this "facilitating the flow of knowledge to the people" is just the supply side of the equation—the technology transfer function. Facilitating the flow of knowledge to the people has very little to do with the people participating in decision making about the agenda of scholarship or enhancing that scholarship.

As our society becomes ever more complex, there emerge new problems to add to an unending list of unsolved problems of our people. Who will set the scholarly agenda, and how do they know that it is valid and relevant, and relevant for whom? In general in the society, except for the broadest categories of science funding from the U.S. Congress, it is scientists who set the scientific scholarly agenda. It is called "peer review." "The . . . ritual that gets in the way of good science is peer review . . . . The term 'peer review' in the context of science policy has acquired a deep symbolism within the science community. It is repeated like a mantra or used as a talisman to shield any activity, put it above reproach, so to speak" (Shapley and Roy 1985). On the other hand, part of what has made for the high returns on investment of agricultural science and extension has been the focus of the research agenda. Just as the payoff to medical research is ultimately in terms of proof in clinical trials, the feedback from farmers through the extension system has helped to define where the cutting edge in agricultural science should be, and where the highest payoff would be.

Can extension programs and other outreach efforts also contribute to the definition and refinement of the scholarly agenda in areas other than agriculture? Of course they can! Indeed, that may be the only thing that will make land-grant universities distinct from the other research universities—from private research universities. It may be the only thing that justifies continued public investments in them and that redefines their social contract and makes them once again "the people's universities."

The land-grant universities were to be better than Harvard, Yale, Cambridge, and Oxford under the norms of 1862 America. Today they may be no different. Just like Harvard and Stanford, they sort through the youth of America to find those most likely to succeed and put their imprimatur or brand name on them. Today, to their credit, the land-grant universities successfully compete with Harvard, Stanford, and MIT for research grants, contracts, and the best students. But like the private institutions, the land-grant universities have virtually no research agenda of their own that is directed to the people of their own state. There is little institutionalized and funded effort that feeds the problems of the society into the university, that lays claim to the intellectual resources of the university, and that participates in setting the scholarly agenda, except the individual choice of scholars to search out public and private grant- and contract-funding sources. There is no institutionalized test of relevance or of workability of much of the science practiced at these institutions.

Returning to Our Roots: The Engaged Institution, the report of the Kellogg Commission on the future of state and land-grant universities discusses their roles in the society beyond their roles in formal instruction. The Kellogg Commission is associated with the National Association of State Universities and Land-Grant Colleges (NASULGC). To say that a "return to roots" is necessary if the universities are to be engaged in the society and their multitudinous problems is harsh language about the current state of public university affairs. This from the Commission made up of presidents and chancellors of many of the leading public research universities in the country, and implicitly from all of them, by virtue of their association with NASULGC. The defining statement by the Kellogg Commission as a preface in the document states:

> In the end, what the bill of particulars adds up to is a perception that, despite the resources and expertise available on our campuses, our institutions are not well organized to bring them to bear on local problems in a coherent way (Kellogg Commission 1998).

In commenting on the work of the Kellogg Commission, C. Peter Magrath, president of NASULGC summed up the dilemma of the land-grant universities at the end of the 20th century:

> Our universities, and therefore our society, face a crisis. Public universities must be financially stable and enjoy public confidence in order to

perform their unique and vital mission as the intellectual and educational service centers for America in the 21st century. But to earn this support they must examine themselves, aided by friendly but not uncritical outside counsel—and then change and reform wherever needed to better serve society (Magrath 1996).

In giving a mandate to the Kellogg Commission, the leadership of the Commission, E. Gordon Gee (chairman), president, The Ohio State University; Dolores Spikes (vice-chairwoman), president, Southern University System; John V. Byrne, (director), president, Oregon State University; C. Peter Magrath, president, NASULGC, all eminent leaders of American higher education, included the following in their joint statement:

> To state the case as succinctly as possible: We are convinced that unless our institutions respond to the challenges and opportunities before them they risk being consigned to a sort of academic Jurassic Park—of great historic interest, fascinating places to visit, but increasingly irrelevant in a world that has passed them by (Kellogg Presidents' Commission 1996).

The extension system is a logical candidate to become the institutionalized mechanism for engagement of the land-grant universities with the people of the state—it has played that role in the past. However, in order for that to happen again into the 21st century, there will need to be significant change and a renegotiation of institutional commitments in the universities, in extension, in state legislatures, with established clients of extension, and with the people of the states. It will require new definitions of scholarship and a new epistemology of science.

Despite the dysfunction in the system, some of which will be described in great detail, there are many exciting things happening within extension and outreach from land-grant universities throughout the country. Some of the hopeful things are happening because of settings that encourage and some in spite of the institutional settings that control the extension/outreach function. A few of the promises and possibilities are chronicled, particularly in Chapter 8. Among those hopeful things described are institutional arrangements that empower or are instructive, others are programs that are the result of heroic efforts by individuals proving that people make institutions perform in the end, and some are clever and interesting techniques that anyone can use to personal advantage.

In the pages that follow, first comes the description and diagnosis of illness, then evidence that there is hope. Finally there is a vision suggesting that no matter how difficult to achieve, there is the possibility for a renewed social contract with the American people, even if that vision is an imagined one.

The original mission of the land-grant university is being renegotiated in some places and abandoned in others. In some places, the renegotiating of the social contract is being lead by extension, and in other places, extension is being left behind, in part because dealing with agricultural interest groups is simply too much trouble, given their growing inability to deliver in the political process. As the land-grant universities move into the 21st century, some of them will be land-grant universities by name and history only, and some will again be people's universities. Some will be "state-supported" universities, some will be "state-assisted" universities, and some will be "state-located" universities. Extension can be part of the problem or part of the solution. Being part of the solution will not be easy but it will be worth it.

Notes

1. Some of the snobbery associated with the aristocratic education persists. In the late 1980s, a soccer team from the University of Massachusetts was playing and beating Harvard's team in Cambridge. In frustration at the beating they were taking, Harvard fans jeered the UMass fans with "you will work for us." Personal conversation with a UMass fan present at the incident.

2. Australia, Austria, Belgium, Canada, Denmark, Finland, France, Germany, Greece, Iceland, Ireland, Italy, Japan, Luxembourg, Netherlands, New Zealand, Norway, Portugal, Spain, Sweden, Switzerland, Turkey, United Kingdom, United States.

3. Phrase used by Jonathan R. Cole with reference to research universities in Balancing Acts: Dilemmas of Choice Facing Research Universities, in *The Research University in a Time of Discontent,* eds. J. Cole, E. Barber, and S. Graubard. The Johns Hopkins University Press, Baltimore and London, 1994. "Academic crown jewels" was used earlier by Castle in reference to land-grant universities in his 1980 Kellogg Foundation Lecture to the National Association of State Universities and Land Grant Colleges, *Agricultural Education and Research—Academic Crown Jewels or Country Cousin?* Resources for the Future, Inc., Washington, March 1981.

4. The Carnegie Foundation Classification of Higher Education groups American colleges and universities according to their missions, principally research and teaching. Within the research mission they have two categories and within the teaching mission they group institutions according to the highest degree they offer: doctoral, master's, baccalaureate, and associate of arts. They also identify a class of specialized institutions and a class of tribal institutions.

Research Universities I: These institutions offer a full range of baccalaureate programs, are committed to graduate education through the doctorate, and give high priority to research. They award 50 or more doctoral degrees each year. In addition, they receive annually $40 million or more in federal support.

Research Universities II: These institutions offer a full range of baccalaureate programs, are committed to graduate education through the doctorate, and give high priority to research. They award 50 or more doctoral degrees each year. In addition, they receive annually between $15.5 million and $40 million in federal support.

Doctoral Universities I: These institutions offer a full range of baccalaureate programs and are committed to graduate education through the doctorate. They award at least 40 doctoral degrees annually in five or more disciplines.

Doctoral Universities II: These institutions offer a full range of baccalaureate programs and are committed to graduate education through the doctorate. They award annually at least 10 doctoral degrees—in three or more disciplines—or 20 or more doctoral degrees in one or more disciplines.
(http://www.carnegiefoundation.org/classification/index.htm)

2

The Land-Grant University Interest in Public Service

Introduction

There is a special calling for higher education in most western societies and that calling has much to do with the often cited "public service mission" of universities. Fulfilling that mission has much to do with being worthy of the special calling awarded the academy and its members—the privileges carry obligations. In addition, administrators of public universities have a practical political interest in public service as they defend university budgets in the face of increasing competition for fewer public dollars. It is this practical political interest that is now addressed.

Though there is some ambiguity in the distinctions among "public service," "outreach," "extension," "extended education," and more recently "engagement," all envision some kind of educational activity involving people who are not registered in degree programs and counted as a part of the university student body. The choice of word or phrase used to describe this activity will depend on who you are and whom you are addressing. You may call it "public service" if you are an administrator conveying to legislators that your institution does more than teach students, or a faculty member speaking to the public about the university programs that serve the community. You may call it either "outreach" or "extended education" if you are an administrator speaking to faculty seeking to urge them to participate in a different kind of teaching activity. If you are within the colleges of agriculture or speaking to agricultural audiences, you will likely call it "extension" or "extension education" because of the long tradition of such audiences with that descriptor of the function.

Regardless of the term used, the efforts so described seek to reach a different audience than the students and parents of students served by normal degree-granting instructional programs. The educational content of such service or outreach efforts may be delivered in formal classroom settings, via university publications or electronic media instruction, by informal on-the-job consultation or conversation with university personnel, or passively through the use or enjoyment of some university facility or activity. This chapter elaborates some ways in which such activities occur within the university and attempts to describe what should or should not be counted as public service.

This chapter also explores the origins of the public service mission of universities, the character of the activities described as such, and the character of the transactions between recipients of services and the university as service provider. Discussion of the incentive system that influences the behavior of academics, mitigating their performance of public service in the name of the university, and the important contributions of public service to science and scholarship is deferred to the discussion of scholarly practice and the character of science in the university in Chapter 3.

The Roots of the Public Service Tradition

There are three major traditions or influences on universities in Western countries, according to Stephens and Roderick (1975). The "English" model of the 19th century was elitist, emphasizing the concept of "liberal education"; it stressed the needs of the individual, the quality of teaching, and the special relationship between tutor and student. So important was the tutorial to the dons of Oxford and Cambridge, as well as their beliefs in the individual tuition not only of a man's mind but of his character as well, that some referred to these universities contemptuously as "finishing schools for gentlemen." They emphasized the arts and humanities from the humanities traditions of the Renaissance and eschewed the practical insights of the new sciences.

In contrast to the English model, the central concept of the German university since the early 19th century was *Wissenschaft,* an empirical approach to all knowledge, along with a concern for the increase and dissemination of knowledge. While the English university tended to be student-oriented and the German subject matter-oriented, there was also an emphasis in German universities of serving the professional needs of the state for trained manpower, since they were state institutions. This concern for "professional" needs was extended to the economic and industrial needs of a technologically-based society.

Scottish universities were more democratic than either the English or German ones, perhaps because their students were poorer and tended to live at home or in lodgings rather than in residential colleges. They emphasized both research and teaching. Further, their acceptance of the sciences and new technologies as scholarly subject matter gave them a crucial role in servicing the manpower needs of industrial Britain.

While both the German and Scottish models put more emphasis on modern knowledge and the needs of the state than did the English schools, the sense of societal obligation for American universities comes out of the populist political movement that gave rise to the American land-grant universities. "the Morrill Land-Grant Act of 1862 produced for American universities a mission of service to society to add to the traditions of teaching and research," according to Sir Fraser Nobel (1979, 407), principal and vice-chancellor of the University of Aberdeen.

It was not until the Smith-Lever Act of 1914 established a "cooperative extension service" at each land-grant institution that the public service function of the land-grant colleges really became institutionalized. Prior to that act, farmers approached scientists at the land-grant colleges for individual assistance to such an extent that some reported difficulty in attending to their normal duties. Indeed, as Rainsford (1972) makes clear, the Smith-Lever Act was passed because the direct benefits to farmers, sought by agricultural interests in their support of both the Morrill Act of 1862 and the Hatch Act of 1887, had not been forthcoming. Most students in the land-grant colleges did not study agriculture, even though they came from farm families; results of research and instruction did not reach farmers because they were not in college but on the farm. Further, the report of President Theodore Roosevelt's Country Life Commission in 1910 called for rural rehabilitation (Kile 1948). In addition, there were well-established models for extension prior to the Smith-Lever Act in the effective work of agricultural demonstration agents, established and led in the South by Seaman Knapp and in the North and West by William J. Spillman (Scott 1970).

The full expression of the service function of the land-grant universities was perhaps best expressed by what became known as the "Wisconsin Idea." This was the belief that the boundaries of the university campus should be the boundaries of the state and beyond and was most particularly articulated by Charles R. Van Hise, the University of Wisconsin president from 1903 to 1918. Van Hise declared that he would "never be content until the beneficent influence of the university reaches every family in the state (Ward 1998, 15)."

The Carnegie Foundation 1966–1967 Annual Report particularly comments on this public service function.

> And because in its day and place this peculiarly American invention in higher education worked, indeed worked brilliantly, it came to have a wide influence on popular notions about the proper "uses" of the university. . . . it was not until the First World War and the period immediately following it that public service began to be regarded as a responsibility of

universities generally. The idea that it was an acceptable function for any academic institution was, of course, given considerable additional recognition as the result of the deep involvement of the universities in the 1942–1945 war effort. Since then, in response to the pressing needs of a maturing society for fundamental solutions to ever more complex problems, public service has become a large and important activity at virtually every university, both public and private, and at many colleges as well (Carnegie Foundation 1967, 9–10).

Full acknowledgement of the public service dimension of universities is found in the comprehensive analysis of the consequences of higher education in America by Howard R. Bowen (1996). In *Investment in Learning—The Individual and Social Value of American Higher Education,* Bowen examines both the monetary and nonmonetary benefits of American higher education. He concludes that the value of the monetary benefits of higher education probably exceeds its cost. He further concludes that the value of the nonmonetary benefits probably exceed the value of the monetary benefits several times over. Major components of the nonmonetary benefits, according to Bowen, are "research and public service." He includes in this category all of the functions or activities of higher education that advance knowledge and the fine arts and that serve the public directly.

Evidence that this uniquely American image of the role for universities has spread can be seen in writings on higher education in Britain. In his 1980 book, *Higher Education for the Future,* Sir Charles Carter discusses the public funding of higher education: "Indeed, the claimant on resources, after the needs of teaching and scholarship are met, should not be a portmanteau concept of 'research,' but rather the public service which institutions which employ considerable brain-power should be expected to give. This is a point more clearly understood in the U.S. than it is here" (1980, 100).

Public service has become a central part of the rationale for public investments in higher education, whether in private or public institutions. It seems reasonable to argue that a corollary is a public expectation of some kind of direct public service, particularly from public universities. Peterson (1975) goes so far as to assert that there is a social contract in this regard and that fulfilling that contract remains a goal for higher education. According to Havelock, however, the university is and has been ambivalent about its role as expert and problem-solver for the practical world. Writing in 1967 about the land-grant origins of the public service role of universities, he asserts:

A century later, however, the image of the U.S. university is not clear even to itself. . . . A struggle goes on between teaching and research interests which virtually crowds out serious consideration of the university's role as the problem-solver and expert for the greater society. Meanwhile, the average citizen looking on from the sidelines insistently asks when the professors are going to stop 'studying' problems and start 'helping' the society by using what they know (Havelock 1969, 3–10, 3–11).

That ambivalence within universities about public service continues today. Public service objectives sometimes are seen as a drain of resources from the central responsibility of the university during a time of severe budgets. Others view them as the essential roles of the university and an institutionalized test of its relevance. Still others see public service objectives as the important means of defending public expenditures for the university with the audience beyond the student body and their parents that are served by outreach activities.

For Taylor, as he is quoted below, the special calling of the university is a "public commission" that has its origins in the Morrill Act.

Frankly and unashamedly, the land-grant charter held that there was no part of human life that is beneath the notice of the university; that there is no positive labor of society that has not its proper dignity. But it held also, beyond this, that thought and action are indivorcible, that the place of the academy is in the world not beyond it, that it is part of the business of the university to demonstrate the connection of knowledge, art, and practice (Taylor 1981, 37).

Public service is thus central to the public commission of the university in society and particularly to the land-grant universities because of their origins and traditions.

Public Service—What Is It?

Public service is viewed from within and without higher education as an integral part of the role of the university, particularly the publicly-funded university. But what is meant by public service? Bowen (1996) places public service under the category of "societal benefits," as distinct from benefits captured by the individual students, who are taught, trained, and given credentials by institutions of higher education.

The public service mandate of the land-grant universities from the Smith-Lever Act of 1914, rewritten in 1953, remains "to aid diffusion among the people of the United States useful and practical information . . . and to encourage the application of the same" (Rasmussen 1989, Appendix D). Within the colleges of agriculture where most Cooperative Extension programs in land-grant universities are administered, the Smith-Lever Act is interpreted to encompass a broad array of subjects that pertain to the problems of individuals, households, businesses, and governments.

Most important about this earliest mandated public service function in American higher education is that it is an active, usually nonformal, functional education activity based on the scholarship of the university and directed to widely dispersed and varied audiences beyond the campus. The Cooperative Extension Service programs of the land-grant institutions still function in this manner in a wide variety of disciplines. Much will be said throughout the remainder of the book about the character of the portfolio of programs within Cooperative Extension, but there is little of that portfolio that is determined by legal constraints of the Smith-Lever Act.

Whether the Cooperative Extension Service should be viewed as a standard for public service in universities, or simply one variant of it, is worthy of attention. In order to develop the basis for a discussion of various characteristics of public service activities, a brief description of some of the most prominent or usual activities that are undertaken in the name of public service will be useful. Bowen (1996) identifies the following as direct public service:

- Services to the public from academic programs that require professional practice through public clinics, student teaching, internships, and student placements.
- Recreational and cultural activities for persons in the surrounding community, (includes dramatic and musical performances; facilities such as gyms, playing fields, golf courses, tennis courts, libraries, and museums; radio and television stations or broadcasts; and spectator sports)
- Programs designed to specifically serve the public—the Cooperative Extension model but includes many efforts not administered by extension and not in land-grant institutions.
- Maintenance of a large pool of specialized faculty talent available for consultation on varied social and technical problems. In some people's views, universities represent the primary source of expert knowledge in all fields, basic and applied.
- Leaders for the administration of programs and missions of local, state, and national government.

In addition, Bowen suggests that there is a significant public service benefit from scholarship that operates primarily in the conservation and interpretation of received knowledge and is "a bulwark of our culture," from scientific research in the natural sciences and in social studies, from philosophical and religious inquiry, from social criticism, from public policy analysis, and from the cultivation of literature and the fine arts. Bowen does not include in his list of direct services the range of continuing education programs and courses that represent a substantial public service effort for many universities. Since these programs are often carried on in formal classroom settings, they represent an intermediate position between the traditional instructional role of the university and the public service role.

Both university policy and public policy in the adult and continuing education area involve questions about the qualifications of entering students and the credentials or certification offered to different types of students. Thus, some courses offered under these programs have no entry requirements and provide no certification, while other programs offer courses that can contribute to advanced degree work or other formal certification of qualifications and have strict formal entry qualifications.

One way of distinguishing between these various public service activities is to describe them in terms that are relevant to the internal resource commitments of the university. Thus the listed public services can be categorized as public service enterprises, public service as an externality or spillover, or public service resulting from the utilization of excess capacity.

Public service activities undertaken explicitly for that reason as a part of the university program fit the "public service enterprise" category. Included is much of the continuing education activities, as well the Cooperative Extension Service. Also included are services to alumni and other specialized public audiences, such as the fans of athletic teams and artistic programs, where explicit investments are made to attract or reach a particular audience. The objectives of the programs in this category usually combine some education or enrichment goals along with goals of institutional maintenance and political support.

Though more complex than this analysis indicates, the large investments in collegiate football and other sports can be understood as a public service program aimed at rallying support to the university from alumni. When viewed in this light, surpluses generated from some of the intercollegiate athletic programs should perhaps logically be applied to enhancing other alumni and public service programs in addition to reducing the university instructional costs of supporting student athletic programs. Further, it also suggests that the athletic programs should be used to rally support from alumni to the total university program rather than just to the athletic programs.

Other public services generated by universities occur as by-products or spillovers—unintentional outcomes—from other ongoing university activities. Clearly the benefits received by the pupils served by student teachers, or by the patients of medical and dental school students, are of this nature.

When research is funded by outside grants or contracts, the results are usually the intended outcome of the funding agency and can only be claimed as public service benefits from the university to the extent that the university has subsidized the direct or overhead costs of the research. The benefits of such research or scholarship must be attributed to the funding organization. It is right to applaud the work of a university scholar who has discovered some new drug to treat AIDS or cancer. But when the work was funded by a drug company or the National Institutes of Health, it is not particularly a "public service" of the university that received the grant but rather a planned and intended outcome of the granting organization. The long-term maintenance of the scholar or researcher in a permanent position that provides career security between research grants is a part of the university's public service.

It is also true that a substantial amount of scholarly work by academics is undertaken without external funding. Results of this activity that contributes to the public, rather than just making it easier for the professor to teach paying students, is a public service. The public service benefits to the people of a particular state from Agricultural Experiment Station-funded research is the intended outcome of that organization and is paid for by the collaboration between the federal government and the state government, through the university.

Some public service benefits generated by universities result from the public having access to facilities or resources that have greater capacity for use than if only members of the university community have access. Obvious examples are museums, art galleries, and some major scientific instruments. Other examples include electron microscopes and, prior to their obsolescence, mainframe computers. Generally this "excess capacity" is because such resources come in lumpy units; in order to obtain the capacity needed for a particular university purpose, you get some amount more than you need. In most cases where public service is the result of the utilization of excess capacity, there is the potential of congestion occurring in the use of the resource, ultimately leading to conflicts between public service uses and campus program uses. The notion that the faculty is a "talent pool" of expertise, the maintenance of which is a public service, can be included within this category.

From this perspective, a variety of issues within the university surrounding faculty consulting activity become primarily questions of who will control and benefit from the utilization of faculty excess capacity—the individual scholar or the university? Some of the conflicts over consulting activity result from disagreements, or lack of clarity, about how much of an individual scholar's capacity is "excess" from the university department's point of view. University policies that are unambiguous on this issue, and that permit both scholar and university to benefit, will likely improve performance for both.

Not only must university leadership consider issues of internal resource allocations when analyzing public service activities, they also need to consider the perspective of the taxpaying voter, who also has some expectation of direct benefits from the university. The question of who pays for the university is relatively important to the university in defining its public service objectives, if those services are intended to elicit support for the university. It is for precisely this reason that the subject of this chapter is about public universities—the private university's political economy of public service is quite different. For publicly-funded universities, the people who pay for it are the "public" who should be served by it.

Establishing just who that public is is not sufficient to guide decisions on what public service should be undertaken when the objective is to elicit public support. Some basis for judging the public response to particular kinds of public service is necessary. The necessary conditions for a public service program to earn and collect credit from clientele are the following:

- Positive Net Benefit Condition: The program must generate a positive net benefit to the client—the total benefits of the education or information must be more than what it costs to get it, including time and travel.
- Attribution Condition: Most of the net benefits, regardless of magnitude, must be attributed to the university.
- The Solicitation Condition: The collection of political capital usually involves a separate transaction. The clients must be identifiable and thus susceptible to being solicited for support.
- The Political Action Condition: Acting politically for the university must cost the clients less than their past and anticipated future benefits. As with all agencies in the public sector, public service activities in land-grant universities do a variety of things to reduce the costs of political action. The widespread use by cooperative extension programs of organized trips for volunteers to Washington as part of the means to influencing federal allocations to extension are just one example (McDowell 1985).

Meeting these conditions is essential to prompt individuals and organizations to make representations to politicians, often a precondition to a receptive attitude when legislation of interest to the university comes before the legislature.

The four necessary conditions also shed some light on the various categories of public service set forth by Boyer and listed above. When the stream of benefits is general, broadly applicable, and simply there for the taking, it is less likely that there will be attribution of specific credit to the university. Such is the case with the use of museums, the talent pool of experts, and the general cultural and intellectual enrichment contributions of the university. Regardless of the method by which the university generates the stream of public service benefits—whether from specific programs, spillovers, or excess capacity—if it is specific to an individual, firm, or organization, they are more likely to attribute it to an action taken by the university. Such is the case with much of the Cooperative Extension program, the benefits obtained as spillovers from various internships, professional experiential learning activities, as well as some specific applied research activities. In a vein similar to the point being made here, Boulding (1975) talks of "visible virtues" as distinct from "invisible virtues" and notes that the marketplace is more responsive to visible virtues—that is certainly also true in the political marketplace.

These four conditions for eliciting political support from public service activity suggest several other insights. Consider the use of user-fees for extension or other university outreach publications or programs. If the publication or program is priced such that all of the value that a consumer would gain from it has been charged in the fee, then there is nothing more to be elicited in the form of political support. It is rather like the advertisement for FRAM oil filters—"you can pay me now, or you can pay me later." You can collect from the audience now, or you can collect from them later. If you have collected all the value at the time you delivered the program, you can't collect again later. On the other hand, some audiences may be so difficult to collect political support from at a later time, that you had better get all you are going to get at the time of program delivery and make your user fee as high as the market will bear.

As a general principle, user fees in public service programs are only nominally for revenue generation. More importantly is the need to accomplish other objectives of the public service program, such as the elimination of trivial requests for assistance. Clearly the highest cost input into any extension or outreach publication is the intellectual input that is embodied in the subject matter content of the publication. Thus, the printing costs or the distribution costs of university outreach publications are nominal costs by any stretch of the imagination. Focusing on "cost recovery" of those production costs is misleading and it is incumbent on public service managers to consider other issues when setting user fees. The arguments are similar to charges for disease immunizations—at some prices you may recover some costs and at other prices you may defeat your public health objectives.

In summary, universities generate public service benefits to the society in several ways—by explicit programmatic design, as the result of spillovers or externalities of other activity, and by providing access to the use of otherwise underutilized resources. Of special interest to university leaders is the political support that can accrue from public service activities. However, not all of the activities called "public service" are equally effective in generating public support for the university. In some cases, the benefits obtained are not attributed to a specific action or intent of the university. In other cases, the public service effects are so diffuse and general that the benefits to an individual are not easily identified or of sufficient magnitude to motivate them to action.

Clearly, the most efficient and desirable public service activity, from the university point of view, is one that can be accomplished with a minimum of additional resources. Further, any benefits that are generated must be experienced by the public in ways that make it easy to attribute them to actions of the university. The benefits to individuals or organizations must be of sufficient magnitude that they will be willing to become advocates or a constituency for the university. In order to keep the costs of public service activities to a minimum, the exploitation of existing excess capacity and potential spillovers from ongoing activities should be considered.

The method and associated cost of undertaking public service activities are not the only considerations in evaluating their usefulness to the university. In addition to an evaluation of the public contribution of a public service activity, are considerations of the support for the university that such activities will generate. Because the packaging of the public service program can affect the Positive Net Benefit condition, the Attribution condition, or the Solicitation condition, the way that public service programs are carried out can be important. Further, unless powerless or disorganized groups in the society have the attention of powerful advocates, there will be a tendency for university outreach to serve well-established and well-connected clientele first.

In light of this analysis, the Cooperative Extension Service with its emphasis on programs delivered through county extension offices is not necessarily the only way to organize public service activities, though even within extension there is considerable variation in the way programs are delivered. Where public service is closely tied to ongoing research and scholarship exploiting both the spillovers from that activity and the excess capacity of the researchers, it is a particularly useful model. Cooperative Extension works extremely well in the domain in which it operates because, in most cases, considerable attention is paid to delivering information to audiences in ways that will make the maximum impact by achieving all of the necessary conditions elaborated above.

However, just as it is appropriate to ask whether football programs as outreach to alumni actively rally support for the whole university, it is also appropriate to ask whether Cooperative Extension programs elicit support for the whole university or only for extension.

Engage Again or Die—The Public Service Bottom Line

The authors of *Returning to Our Roots: The Engaged Institution* (Kellogg Commission 1998) argue that "engagement" is more than public service or extension. Engagement, they assert, means "institutions that have redesigned their teaching, research, and extension and service functions to become even more sympathetically and productively involved with their communities, however community is defined" (Kellogg Commission 1998, vi). The words "more than public service or extension" respond to a view of outreach and extension as essentially a one-way street from the university to the society. As will be argued in the next chapter, the unidirectional flow model has seldom been the view of those most engaged in extension. The distinction between that unidirectional model and engagement does serve to set the context of the arguments about the bottom line of public service as defined above and the interests of the land-grant universities.

Derek Bok (1990), former president of Harvard University, argues persuasively in his book, *Universities and the Future of America*, that America's research universities have much to contribute to America in the 21st century. He explicitly identifies the contributions to greater competitiveness, to a search for a better society, and to moral education. Bok is concerned that without the contributions of the universities, society is in peril. He suggests ways and means whereby universities can organize themselves to engage the world and not "succumb to its blandishments, its distractions, its corrupting entanglements . . . " diminishing the ". . . more profound obligation that every institution of learning owes to civilization to renew its culture, interpret its past, and expand our understanding of the human condition" (Bok 1990, 103–104).

By contrast to Bok's concerns, the thesis of this book is perhaps more mundane and pragmatic. It is that, without greater engagement of the universities with the society, the public universities are in peril. Without that engagement, not only will the universities not be able to contribute to the pressing problems of the society, they will not be able to, in Bok's terms, understand or renew the evolving culture, accurately or effectively interpret history, or significantly expand the understanding of the human condition.

In his announcement of the establishment of the Kellogg Commission, C. Peter McGrath, president of NASULGC reflects on the Bok perspective of new and better engagement of the universities for the sake of the society:

> Yesterday's good works are inadequate for tomorrow's needs. We must recognize the new realities of diminished public resources—while facing our shortcomings forthrightly. Clearly, these include our need to use faculty time more productively, our obligation to pay more attention to undergraduate students and to become full-time collaborators with public schools, and our duty to link research discoveries and educational insights with our states and communities in partnerships that strengthen our economy and society (Magrath 1996).

However, the leadership of the Kellogg Commission is of the opinion that unless there is significant change in the ways that the public universities (state and land-grant universities) conduct themselves they are likely to become irrelevant to the society—"consigned to a sort of academic Jurassic Park—of great historic interest, fascinating places to visit, but increasingly irrelevant in a world that has passed them by" (Kellogg Presidents' Commission 1996). Curiously, "dinosaur" is the descriptor used by concerned extension staff when describing their frustration with the character of the existing extension portfolio and the reactions of colleagues when undertaking programs in nontraditional areas of extension.[1] It appears that some extension field staff are reading from the same page as these renowned leaders of public higher education.

The authors of the Kellogg Commission report advocating greater engagement of the university appear to share this writer's view that the most immediate peril in the detachment of the university from the society is first to the university, and secondarily to the society. Given the history of the land-grant universities, to fail to reengage the society in the face of this peril would be an abandonment of the social contract with the people of America.

Notes

1. Traditional Cooperative Extension programs are 1) agricultural programs primarily directed toward farm production, 2) 4-H youth programs, and 3) home economics programs that focus on traditional homemaking skills and activities.

3

The Academy, Science, and Service

Introduction

"Teaching, research, and service to community dominate the professional lives of men and women in higher education," says the National Education Association in its 1984 Almanac (NEA 1984). In an article intended to assist the professional development of college and university professors, "service" is once again used to evoke the sense of a special calling for academics and the academy. After this initial invocation of service to community as a noble part of professional practice, nothing else is said of it in the article. That treatment of the service function of the university is fairly representative of the attention that is given to it by academics and the academy generally, despite the substantial self-interest that the university has in using public service and extension to establish itself before its public, as argued in the preceding chapter.

Defense of the public service function of the university is, for the most part, only vigorously argued by those few actively engaged in it and by university administrators. This short shrift is a reasonable representation of the attention given to public service and extension in the prevailing culture of the academy generally, including within land-grant universities. Even scholars studying the return to investments in agricultural research and extension, who have included extension in their studies because it is difficult if not impossible to measure and analyze the separate functions, have virtually nothing to say about extension when discussing their findings (Alston and Pardey 1996, Evenson and Kislev 1975, Huffman and Evenson 1993).

All land-grant universities speak in glowing terms of their commitments to instruction and service, and indeed some of them may carry out those functions better than it is done anywhere else in the society. However, the scholarship of discovery—usually called research—is the dominant coin of the academic realm, and it is thus in land-grant universities just as it is in Harvard, Oxford, or Humbolt universities. New knowledge is the main business of the contemporary land-grant university. The fact that at the end of the 20th century, most land-grants are research universities and more than one-third of the research universities of the nation are land-grant universities is prima facie evidence for this assertion.

If this book's discussion of the future of the land-grant universities and extension is to be complete, it must attend to the relationship between public service/extension and research. It must attend to the incentive system that influences the behavior of academics, mitigating against their performance of public service on behalf of the university. It is to these tasks that we now turn.

Public Service—Contributions to Scholars and Scholarship

It is quite common for academics to argue that "research is essential to successful and effective teaching." It is uncommon for them to argue that teaching makes significant contributions to research or scholarship. Given the incentive structures and conventional measurements of performance in teaching and research, the arguments by some professors about the contributions of research to teaching seem at best self-serving, and at worst, an excuse for disinvesting in teaching in favor of research. Though oft claimed, scholarship on the question finds that engagement in research and research productivity is barely correlated with student evaluations of teaching effectiveness with a positive correlation in the range of 0.13 across multiple studies (Feldman 1987). The positive sign on the correlation is reassuring because it affirms that research activity by the professor does not do the students harm, but it does not say much more than that. So much for the relationship between teaching and research.

In earlier sections, the ways in which scholars, scholarship, and the university generally contribute to public service have been described. But what about the contributions of public service to research or discovery scholarship? In this section that relationship will be explored.

Better Science

According to Blaug (1980) there has been great turmoil among those who have philosophized about science and scientific method since the 1960s. Among those challenging previously received theories of science are Sir Karl Popper and Thomas S. Kuhn. Both Popper and Kuhn agree that most scientific advancement does not come about primarily by accretion but by the revolutionary overthrow of an accepted theory and its replacement by a better one. However, they disagree substantially on whether the day-to-day work of scientists is revolutionary or not, on when scientific tests are challenges of theory or tests of the ability of the scientist, and on whether applications of science that are less than a test of fundamental theory are "hack" science or a necessary condition to generating revolutionary changes. Neither Popper nor Kuhn believes in induction as valid scientific method since there are no rules for inducing correct theories from facts—there is no logical basis for "validation." Rather, both believe that the "falsification"—Popper's term—necessary to advancing knowledge is only possible from deductive reasoning—from hypothesizing, testing, and rejecting.

Kuhn (1970) asserts that the fundamental issue on which he and Popper agree is that an analysis of the development of scientific knowledge must take into account the way that science is actually practiced. Based on this insight about the importance of the behavior of scientists in the practice of their craft, the argument is made that the engagement of scientists in solving real, practical problems via an involvement in public service activities, directly or indirectly, contributes to the advancement of discovery scholarship and perhaps even the solving of theoretical problems.

Both Kuhn and Popper emphasize that it is through the deductive process of repeated testing of scientific theories and the associated rejection or failure to reject that scientific advances are made. Clearly the laboratory and experimental conditions and procedures prescribed by statistical analysis and the various scientific disciplines provide the most rigorous conditions for Popper's falsification. Kuhn, however, argues that Popper's emphasis on falsification in the advancement of knowledge gives too much emphasis to unusual and extraordinary research, and too little emphasis to the day-to-day work in the practice of science. It is this work, which is mostly solving puzzles rather than testing hypotheses, argues Kuhn (1970), that hones the skill of the scientists such that on some occasions scientists actually are able to set forth hypotheses and perform experiments that test fundamental theories and advance scientific revolutions.

In describing his disagreement with Sir Karl Popper on scientific practice and the importance of solving puzzles, Kuhn writes:

> It is important to notice that when I describe the scientist as a puzzle-solver and Sir Karl describes him as a problem-solver, the similarity of our terms disguises a fundamental divergence. Sir Karl writes (the italics are his), "Admittedly, our expectations, and thus our theories, may precede, historically, even our problems. *Yet science starts only with problems.* Problems crop up especially when we are disappointed in our expectations, or when our theories involve us in difficulties, in contradictions." I use the term

"puzzle" in order to emphasize that the difficulties that *ordinarily* confront even the very best scientists are, like crossword puzzles or chess puzzles, challenges only to his ingenuity. *He* is in difficulty, not current theory (Kuhn 1970, 5).

Kuhn further emphasized the importance of solving puzzles, in contrast to testing theories in scientific practice when discussing the practice of astrology. He said that astrology cannot be dismissed as unscientific on the basis of the vague and imprecise way that its practitioners couched their predictions making refutation difficult, on the way that they explained its failures, or even on the basis of its limited success in prediction. Many of the same criticisms, he suggests, could have been levied at engineering, meteorology, and medicine more than a century ago. Each of these respective fields, which was at the time more akin to craft than to a science, had shared theories and craft-rules, which guided practice and established the plausibility of the discipline. And while there was great desire for more powerful rules and more articulate theories, it would have been absurd to have abandoned their practice simply because the desired new insights were not at hand. In the absence of a new set of rules of practice, neither medicine nor astrology could carry out research. " . . . they had no puzzles to solve and therefore no science to practice" (Kuhn 1970, 9).

In comparing the early practice of astrology with that of astronomy, often practiced by the same people, Kuhn makes the point that while individual failures in prediction in astronomy would give rise to a host of calculation and instrumentation puzzles, the same was not true of astrology. There were too many possible sources of difficulty, most beyond the control of the astrologer. Thus, while individual failures could be explained, no one, no matter how skilled, could make use of them in a constructive way to revise the astrological traditions. "And without puzzles, able first to challenge and then to attest the ingenuity of the individual practitioner, astrology could not have become a science even if the stars had, in fact, controlled human destiny" (Kuhn 1970, 9–10).

Johnson and Zerby (1973) speak of the distinction between practical and theoretical or disciplinary problems when discussing the way in which economists deal with values, because, they assert, it is impossible to address human problems without reference to values. The solution to practical problems—perhaps more akin to Kuhn's puzzles—they argue result in action, and demand resolution. Theoretical or disciplinary problems are often never resolved—apparently consistent with Kuhn's disagreement with Popper that the testing of theories is not the usual, day-in, day-out work of scientists and are rare events.

In order to deal with values in the process of solving practical problems, Johnson and Zerby (1973) argue that scientists must engage both practical and theoretical beliefs. Practical beliefs can be either descriptive or prescriptive. They are beliefs about the nature of reality, both normative reality (what people believe) and nonnormative reality (what is), and about the rightness and wrongness of possible solutions to the practical problem at hand. Practical descriptive beliefs, whether about normative or nonnormative reality, are only of practical value, they argue, when combined with prescriptive theoretical knowledge to yield descriptive prescriptive knowledge. Theoretical or disciplinary problems involve beliefs about whether alternative normative and nonnormative concepts describe reality. In finding solutions to practical problems, it is necessary, argue Johnson and Zerby (1973), to use theoretical, nonprescriptive beliefs about both normative and nonnormative reality.

After reemphasizing that practical problems cannot be solved without reference to theoretical questions, Johnson and Zerby (1973) continue the discussion of problem solving by pointing out that the application of knowledge in solving problems is a creative enterprise requiring objectivity. "Objectivity" is used to describe both the investigator and the kind of knowledge that results from objective investigation. The investigator is considered objective when she refrains from identifying herself and her prestige with a particular concept, and will thus be willing to submit the concept to various tests of objectivity. Knowledge or concepts are objective when they pass tests based on rules of evidence and valid means of justification.

It is incorrect, assert Johnson and Zerby, to say that a statement is objective because it is true, or even that the statement is objective because it is an accurate description of reality. The latter implies that our experience tells us when there is a correspondence with reality—the only check on that is more experience, which may be as flawed as the first.

A concept is objective, suggest Johnson and Zerby (1973), if it

- "is not inconsistent with other previously accepted concepts, and with new concepts based on current experience,
- has a clear and specifiable meaning, and
- is useful in solving the problems with which one is confronted" (Johnson and Zerby 1973, 224).

The first test of objectivity—the test of consistency—includes both *internal consistency* and *external consistency*. Internal consistency is an analytical test and requires concepts to bear a logical relationship with each other. The advantage of mathematical models as representations of theoretical knowledge is that they are, by definition, internally consistent. However, there are sometimes problems with such models passing the test of external consistency. The test of external consistency is a test of experience based both on synthetic knowledge (derived from experience) and analytic knowledge (deduced by logic from propositions). New or independent experiences can be derived through observation such as is accomplished by statistically designed experiments, survey research, or other approaches to observation. Observations or experience provide a basis for forming new concepts. To apply the test of external consistency, the newly synthesized concept is analytically compared with existing concepts.

The test of clarity, the second test of objectivity, is simply the meaning of clarity. If a concept can be easily articulated and communicated, then it will pass the test of clarity. If not, then not.

The third test of objectivity in practical problem solving is the test of workability. It is a test that comes from pragmatism, which, argue Johnson and Zerby (1973), is primarily interested in the usefulness of knowledge. They illustrate the workability test by suggesting that the assumption that light moves in a straight line passes the workability test of objectivity if the problem being solved is the sighting of a rifle. Presumably, if either interstellar travel or molecular behavior is being contemplated, then quantum insights to the behavior of light must be considered to pass the test of workability. Similarly, the assumption that the earth is flat is "workable" when contemplating the construction of a building or a bridge, but not when plotting intercontinental air routes. In order to site the house for the best passive solar heating in the northern hemisphere, pure empirical observation will lead one to choose a southern exposure. In order to explain the empirical results, one will likely have to abandon the flat earth assumption. It is particularly in learning about the circumstances where this test of workability is met that scholars gain from attempting applications of their science to real world circumstances.

There are two points to be made from this formal discussion of the way that science is practiced and scientists behave. One is that, contrary to Popper, the daily practice of science in less than "falsification" of basic theories but rather in applications or practical problem solving is not "hack" science, but a different brand of scholarship. The second point is that this type of scientific practice has as much likelihood of contributing to Popper's "falsification" and ultimately to scientific revolutions, by virtue of the fact that in the solving of practical problems or puzzles, the scientist has some external discipline impelling a decision.

Perhaps there will be someone who will act on the proposed solution and will carry out an additional institutionalized test of workability.

By engaging in such problem-solving activity, the skill of the scientist is increased and she may, in Kuhn's terms, be more likely to be able to set up the experiment that actually tests the theoretical hypothesis. The test of objectivity that permits the scientist to work from her discipline on the practical problem is the test of consistency. The test that permits an actual solution to be found to the practical problem at hand is the pragmatic test of workability—light moves in a straight line or is influenced by gravity depending on the application.

To restate the point: the exposure of the scientist and her theories to the rigors of application in a practical problem ("puzzle" in Kuhn's terms) not of her choosing provides a clear test of the capacity and knowledge of the scholar, and perhaps also a test of the validity of the theory.

Relevant Science and Relevant Universities

The distinction between "scholar" and "researcher" made by Carter (1980) is useful here. It is similar to the concerns of Popper over the danger of specialization. Carter understands scholarship to be broadening, integrative, and extensive. Though the scholar may at some points have reached the "frontiers of knowledge," it is not essential that he himself be pushing those frontiers out. On the other hand, according to Carter, the discoverer—his term for researcher—is in danger of becoming a person who knows more and more about less and less, since he is "working on a tiny section of a long circumference" (Carter 1980, 97), given the everexpanding knowledge base of many disciplines. Moving out of the ivory tower and exposing the academic to problems in the real world provides a potential for a real contribution to scholarship, and perhaps even directly to research. It is the question of relevance. At first, this may seem to be a restatement of the "test of workability" argument made above. However, the question of the relevance of scholarship is broader, and is frequently discussed in the context of the recruitment of students to a discipline or the relevance of the discipline to contemporary problems.

According to Feld (1975), there had been a sustained decline in the proportions of students willing to commit themselves to careers in pure science over the previous 50 years. This trend was particularly strong in the period from 1970–1975. Shapley and Roy (1985) give evidence that the trend was continuous until 1985. Feld, a professor of physics at MIT, attributed at least part of this decline to "the desires of young people for involvement in pursuits 'relevant' to human concerns, coupled with the image of pure science as being aloof from, if not hostile to, societal needs" (Feld 1975, 244). The solution advocated by Feld was greater involvement of pure scientists in public service. The title of his paper, Legitimizing Public Service Science, and its discussion speak to some difficulty in achieving that among basic scientists and describes the price the sciences pay as a result.

In the Twelfth Congress of the Universities of the Commonwealth, held in August 1978, the theme of "relevance" was a prominent one, as scholars and leaders of Commonwealth universities discussed the pressures they had to deal with. Sir Charles Wilson, principal emeritus of the University of Glasgow, was somewhat hostile to the notion of a need for greater relevance in teaching and research as evidenced by greater responsiveness to local, national, and international problems. His objection was that "those who would make a whole philosophy out of 'relevance' would . . . like the universities to come closer to the world of action and practice and to sacrifice some of their detachment in favor of social involvement" (Wilson 1979, 22–23). For Wilson, the risks associated with involvement in daily events are a loss of scholarly detachment and neutrality. The 1978 concerns of Wilson sound remarkably like Bok's 1990 concerns when Bok argues that engaging the world may be a distraction to the "more profound obligation that every institution of learning owes to civilization to renew its culture, interpret its past, and expand our understanding of the human condition" (Bok 1990, 104).

John F.A. Taylor (1981), in his book *The Public Commission of the University*, poses an answer to the Wilson and Bok concerns.

> To perform its public office and to do its proper work, the university must preserve itself beyond partisanship and beyond advocacy. The university nails no theses on church doors. Its proper work is done when it establishes the public conditions of rational exchange, when it institutes the convocation in which partisanships may be impartially heard, the collisions of opinion peaceably resolved, according to rules known and commonly admitted in advance (Taylor 1981, 17). . . .
>
> Socrates used to say with disarming civility: "Let the argument lead us." That is what the neutral university says. The argument is independent of the disputant; it must be kept permanently in the public domain. The business of keeping it there is the university's public commission (Taylor 1981, 19). . . .
>
> Of this, only are we perfectly assured that in the new relation of science and society there can be no such thing as a university beyond politics. A mere silence on public questions will not prove its innocence; quarantine will not prove its loyalty. The path of a university is unavoidably a political path for the reason that neutrality is unavoidably a political role. The

problem of neutrality is not how to be out of the world but how to be in it—how to be in it without being of it (Taylor 1981, 29).

According to this notion of the public commission of the university, part of scholarship is to be aware of societal issues related to the particular area of scholarship—to be relevant. The test of relevance affects both the agenda of the scholar and the conduct of the scholarship. To ensure that this relevance in scholarship occurs is important to the administration of the university since, as with academic freedom, it is not important just for the sake of the scholar but for the sake of the university and the society.

For many academics, the exposure to real world problems comes through consulting activities rather than through public service. Indeed, consulting, like public service, makes a positive contribution to scholarship through both the "test of workability" and the "test of relevance." However, it is important to understand the direction in which the flow of benefits is moving and not to confuse this benefit to scholars from consulting with the activity and concept of public service.

This discussion of relevance also provides some basis for comment on the difference of perspective between Bok and the Kellogg Commission on the Future of State and Land-Grant Universities about greater engagement of universities in the society. Bok is concerned that without engagement the society is in peril, whereas the Kellogg Commission is concerned that without engagement with the society the public universities are in peril. There is an implicit concern by the Kellogg Commission that without engagement the universities will have nothing to say to the society and that will be a social loss. Ensuring the involvement of university scholars in public service becomes an institutionalized test of relevance.

The Bok view clearly comes out of the private, elite university culture and its remnant aristocratic view of the university and society. The public university/land-grant university leaders more closely reflect the populist, public support/public obligation perspective that is part of the social contract between the American people and the land-grant universities.

Taylor (1966) provides perhaps a balance between Bok and the Kellogg Commission and the possibility that the university would become only an instrumental agency of the society:

> He who regards the university as an island, who lets it become one, is treasonable to it. He provincializes its community and diminishes himself. He gains a province and loses the world. He may even gain the world, but he shall have denied its soul (Taylor 1966, 225).

In its relation to society the university's function is, in the first instance, to provide the means to ends that society has chosen for itself. But it is a lame architect who houses an activity without civilizing it. You do not sensitively house the life of a man by providing only for the movement of his bowels, and if in seeking to serve his needs you search out only the known needs which he declares and will think to define, that he needs a kitchen and a place to lay his head, you will serve him very ill indeed. He buys the services of an architect; you give him the services of a privy-carpenter (Taylor 1966, 228).

Public Service and the Academic Community

It has been argued in this chapter and the preceding one that public service/extension is an important university contribution to society; that it is of considerable value to university administrators as they make a claim for public support; and that it is important to the quality and relevance of the scholarship, including research, undertaken at universities. In this section, it will be argued that despite these highly valued contributions, there are significant disincentives within the academy, which mitigate against the performance of public service activity by most academics, even in land-grant universities. Further, it will be argued that any effort to enhance the public service output of a university must take into account the incentive system that is the dominant influence on all activity within universities.

Academia's Misunderstanding of Tenure

The covenants within society that sustain a community of scholars with the privileges academics have in Western countries are discussed at some length by Taylor (1981). According to Taylor, the notion of the scholar as a "delegated intellect" whose academic freedom is defended by both the employing institution and the courts, and who is provided with tenure in that employment, is based on the public perception that "the work of thought is public business, and the shelter of argument a public trust" (Taylor 1981, 23). This public commission of the university and of scholars, according to Taylor, is independent of the financial circumstances of any particular institution, and applies equally to private and public universities.

The legitimate defense of academic freedom, of which tenure is but one mechanism, is based on the public need to hear what scholars have to say, rather than with a concern for an employment perquisite for professors as one class of employees of universities. Taylor observes that the relationship between university trustees and academics is not one of "employment," rather it is similar to the tenured appointment of justices on the U.S. Supreme Court, and for much the same reason. Tenure of Supreme Court judges is not for their sake but for the society's.

Unfortunately, at the end of the 20th century, the public perception of tenure for academics is that it is assured employment and has little functional benefit to the society. That is perhaps an earned perception of academics and their tenure because too many have failed to meet the obligations of tenure. Among academics, the discussions of tenure are almost always about the importance of tenure to academic freedom and about tenure as a perk of the job. Almost never is there a discussion of the obligations of tenure—of the obligation to speak to a social issue out of one's expertise, because one is protected by tenure.

Recent efforts to do away with tenure in universities on the grounds that it is dysfunctional, such as occurred at the University of Minnesota in 1997, are likely the result of both this public perception of tenure, as well as a generally growing distrust of the integrity of scholars and science. Clearly, the serious side of the science fiction thriller, *Jurassic Park* (Crichton 1990) is of science and scientists run amok for profit.

However, the need for, and the misunderstanding of, tenure persists in land-grant universities in 2000. Dr. Lawrence Cross, a professor of educational research and evaluation at Virginia Tech, who has devoted much of his career to the field of educational measurement, undertook the obligations he felt to speak out on a public issue related to his field. In Virginia, state government has undertaken a major effort to reform Virginia schooling with leadership from the governor and the state board of education, and have established a set of tests to measure the Virginia "Standards of Learning" (SOL). The SOLs were originally drafted to represent guidelines for what students should learn and what teachers should teach. Now, however, students, teachers, and schools are being held accountable for performance on tests, first implemented in 1998, to measure the knowledge and skills specified by the SOLs. The stakes are very high since school accreditation and students' acquisition of a school diploma are to be based on the test scores.

The advocates and opponents of using the SOL test scores as the basis of school accountability have been split along political party lines, with Republicans the initiators of the program and Democrats in opposition—a public policy buzz saw. Cross was concerned that the brief, multiple-choice SOL tests were inadequate to support the use of the test scores in the accountability program and had written extensively on the subject in both professional settings and in public information pieces.

Because the state legislature was debating the use of the SOL tests in its 2000 session, Cross felt that he had an obligation to share with the members of the legislature his professional concerns based on a career of scholarship on the subject—the obligations of tenure. He prepared a synoptic document entitled "Myths and Facts Regarding the SOL Reform Movement" (Cross 2000) and sent it with a cover letter to each state legislator in January 2000. Some, whose views it affirmed, wrote appreciatively of his contribution. However, others in the legislature whose views are not affirmed by Dr. Cross's information will surely complain, and, it is rumored, have complained, to university leadership.

To date, Dr. Cross has not heard directly about his actions. It is clear, he would never have even participated in the public policy discussion of the use and abuse of the Standards of Learning without his tenure being assured (Cross 2000a).

Scholarly Communities

Taylor (1981) also makes clear that the important divisions within the community of scholars is not to be found in loyalty to Harvard or Michigan State, but in the subcommunities of the several scholarly disciplines—of physics, economics, and botany. The foundations of these communities are based on a covenant of method and consent to rules of discussion, proof, and evidence. Anyone, anywhere, who agrees to the rules of scholarship in a particular academic field can participate in the community and its discussions. These communities are supranational.

> Scattered among the nations of the earth, scholars are like the Jews of the dispersion. They have community by covenant, not by fact and by affirmation, not by neighborhood. They are bound to each other in one community in spite of all the political and moral estrangements that separate them in other connections and for other purposes, from one another (Taylor 1981, 21).

The official currency within these communities is the written word. Most of the members view themselves as independent professionals, similar to doctors and lawyers in private practice, who are more responsible and responsive to colleagues in other universities and even other countries than to deans, chancellors, or presidents. The appropriate image of a university is not of a department store with many different departments and different products under one roof, and central management. The more accurate image is of an open market where individual entrepreneurs erect their own stalls and conduct their business pretty much independently of others. The shoe sellers are generally together, as are the sellers of kitchenwares, and all are minimally responsive to those who provide the roof, electricity, water, and sewer. Clark Kerr, former chancellor of the University of California, is reputed to have described the university as a group of professors with a common parking problem.

Taylor (1981) helps us to understand the public context of the university and the community of scholars. He provides insight into the glue that binds these two important institutions within our society. His insights are also useful in identifying how scholars and universities have distorted their public commission—have confused neutrality with seclusion, relevancy with advocacy, and scholarly obligations with conditions of employment. However, Taylor does not describe the sociology, culture, or politics of the scholarly communities, nor the ways in which they influence the behavior of scholars. In order to understand the lack of response of most academics to the obligation for public service, it will be useful to examine these other dimensions of their community.

In discussing the political economy of scholarship and the incentive structures to which academics are responsive, it is useful to suggest an admirable or ideal, yet plausible, set of personal motives of scholars. The arguments for this case will be a "best case" or upper limit on the influence of the incentive system described. In this context, then, most academics are presumed to be dedicated scholars who wish to advance the frontiers of knowledge in their chosen fields. Most want access to research and other resources that give them the maximum opportunity to pursue their scholarship. Most are driven more by a desire to achieve and contribute than by the promise of fortune. Obviously, the resource requirements for scholarship vary greatly by discipline and within disciplines.

The Realm and the Coin. One of the main requisites to the pursuit of research resources is "scholarly reputation." Although there is some circularity in this argument—the need for resources to advance one's scholarship requires a reputation as a scholar—it is a realistic picture of the problem for young academics and of the "publish or perish" mythology. Scholarly reputation is substantially influenced by the disciplinary community at large, through the control of access to the communication network of each discipline—journals, presented papers, awards, and other such anointing from the community. In this vein, Garvin (1980) writes about the "market for prestige" for both individual scholars and for academic departments.

Many in the academic community believe that the leading lights within a discipline, those who adjudicate access to the journals and to power roles within the community, are so empowered because they are the best minds in the field and have earned the role of high priest. For others, like Brian Martin (1981) in his article, "The Scientific Straightjacket," the high priests or elite of the scientific community are as subject to the desire for personal aggrandizement and to political misbehavior as any politician in any community.

In the control of both access to grant and contract resources and to publication of scientific results, one procedure dominates the academic and scientific community—peer review. Shapley and Roy (1985) write that there are three rituals that stand between a scientist and fulfilling the drive to do the best science possible. The three are 1) the distractions of "petty tasks," 2) peer review, and 3) the demand for excessive publication. Peer review, say Shapley and Roy, has taken on almost religious or talisman meaning, shielding much of science from external scrutiny (Shapley and Roy 1985, 103). That peer review may be less noble or indeed an instrument of power within scientific communities, and thus not worthy of the accord given it is increasingly evident as the practice of science is more closely scrutinized.

Toward the end of their book, *Peerless Science*, Chubin and Hackett (1990) state:

> Scientists are at the mercy of peer review systems that may offer neither "peers" nor "review." Instead, applicants must compete with others' intellectual capital, positional advantage, and political clout. Luck of the draw or mere chance may matter nearly as much as measurable features of the manuscript or proposal. Under current conditions of high competition for research funds and space in first-rate journals, such nonmeritocratic criteria make a decisive difference at the margin. Transcendently brilliant science will generally be funded or published and arrant nonsense usually will be turned away, but between those extremes, chance and its less respectable relatives will play important roles in allocation decisions (Chubin and Hackett 1990, 194).

By way of an example of political control in a scholarly community, Homa Katouzian (1980) in *Ideology and Method in Economics*, argues that the tendency for the economics profession to view mathematical formulations as inherently superior to those that use other methods, is irrational. It persists, he says, substantially as a result of the dominance of the profession by mathematical economists and of their inability to see the logical inconsistencies in the mores or ideology they perpetuate within the economics community (Katouzian 1980, 167–168). Feyerabend (1978) cites yet other examples:

> Acupuncture, for example, was condemned not because anyone had examined it, but simply because some vague idea of it did not fit into the general ideology of medical science or, to call things by their proper name, because it was a "pagan" subject. (The hope for financial rewards has in the meantime led to a considerable change of attitude, however.) (Feyerabend 1978, 135)

Whether the view is of bad politics or good politics, it is useful to remember that the process of scholarly recognition and the development of reputation are subject implicitly, and even explicitly, to a political process. Thus it is that a particular scholarly insight can be much more than the solution to a sticky intellectual problem. It can simultaneously be heresy to the received orthodoxy of the discipline, and a major threat to the self-image and perhaps even careers of others in the community—if you can get it published.

Aside from seeking to build a scholarly reputation, the scholar employed as an academic is interested in improving her employment circumstances. Indeed, while the disciplinary community at large does much of the anointing and approving of the scholar, it is the employing institution that gives monetary rewards and employment security—promotion and tenure. Control of that system is also substantially in the hands of the disciplinary community, but in this case, control is in the hands of the department in the particular college or university.

Some academic administrators would challenge the perception that control of faculty evaluation within the university is in the hands of the subject matter departments. It is safe to say that academic administrators may on occasion be more critical than disciplinary colleagues of a faculty member's work. When that occurs it is often on the grounds of some interest to the institution. However, it is difficult for administrators to approve what the department does not approve, or to reward beyond the department's recommendation. Indeed, when this happens, administrators are at risk of being accused of political chicanery, even in cases in which truly perverse political behavior within a department or college has unfairly denigrated the work of a particular faculty member. Some brave administrators do on occasion approve tenure counter to department recommendations, but they do so at considerable risk.

In thinking about the academic department as the locus of scholarly evaluation within the university, it is important to remember that the diversity of viewpoints and subdisciplines represented in the larger scholarly community are, in some degree, also represented within each department. Indeed, many departments explicitly choose faculty to broaden the disciplinary representation within the department. The pecking order within the larger disciplinary community, which may make the study of elementary particles more prestigious in physics than the more applied solid state physics, also exists within the department but in a modified form, depending on the local political situation. This means that in many departments, individual scholars may have very little in common with other members of the same department, even though they are ostensibly within the same discipline. For many academics, the most meaningful collegial relationships are not with individuals in the department but elsewhere throughout the larger disciplinary community.

Most academic departments give the appearance of placing a high value on the evidence of scholarship—the publication of journal articles and the presentation of scholarly papers. Indeed, the "perish" part of the academic creed is usually envisioned as the denial of promotion and, most importantly, tenure, by virtue of a scholarly evaluation that is mostly in the hands of the department. This perception persists despite evidence from researchers such as Lionel S. Lewis (1975) and a multitude of case examples about academic dismissals that scholarly performance as measured by publication record may be necessary, but is clearly not sufficient for advancement as an academic (Dixon 1976, Martin 1981). Given the diversity of scholarly interests represented within a department and the subject matter pecking order within the discipline, Lewis' findings are not surprising.

Lewis' work is particularly interesting because it makes clear that, at least at the time of his writing, the major other elements in the evaluation of academics was not their contributions through teaching or public service but rather their adherence in personality and behavior to what he calls "a puritan ethic" and "a social ethic." The puritan ethic holds that "self-discipline, austerity, and hard work are the keystones of success" (Lewis 1975, 77). The social ethic "turns on the belief that charm, a conforming personality, or skill in interaction is essential for those who would effectively advance the work of the world" (Lewis 1975, 77).

Another aspect of the evaluation of academics and of scholarship discussed by Lewis is of considerable interest to the theme of this book. According to Lewis, most of the evaluation of actual published work of academics by their fellows is accomplished without actually reading the work. A variety of surrogates for actual review are used, most notably the reputation of the journal in which the article is published. Another widely employed shortcut is a quick perusal of the list of references for recent or known citations. Not only do these practices, which are an accepted convention, reinforce the dominant influence of the larger scholarly community, they inhibit any activity that requires a more substantive evaluation, no matter how meritorious or useful it may be to the department or university.

This in part explains the relative evaluation of scholars whose main investment and contribution is in teaching or public service. Even if their departmental colleagues are supportive of the work, because its proper evaluation is more difficult, it will likely be undervalued, if counted at all. It should be pointed out that the full evaluation of a research or scholarly contribution—without the evaluative convention that accepts the reputation of other scholars and journals at the department level—is more difficult than the evaluation of most teaching and public service activity.

It is in precisely this context that the argument of supranational disciplinary communities maintaining control over both the realm of scholarship and the coin of the realm exists. For the most part, those communities are mute on the question of public service activities by their membership, if not actually hostile to it. The long-term debate about the relative evaluation and rewarding of scholars with major extension assignments within the colleges of agriculture are familiar to anyone with any association with land-grant universities.

There have been a variety of institutional responses in a number of the land-grant colleges of agriculture as faculty with major extension appointments experienced difficulty meeting the tenure and promotion requirements. In an effort to make some of these issues explicit, this writer presented a paper to the 1987 American Agricultural Economics Association, Pre-Conference Extension Workshop. The name given to the workshop was "Maintaining the Cutting Edge" indicating that the workshop would assist those agricultural economists with extension appointments to renew themselves and acquire some of the skills necessary to be closer to the cutting edge of the profession. The paper submitted and presented was "Why Many Extension Economists Are Not at the Cutting Edge and What They Can Do About Moving the Edge" (McDowell 1987).

Scholarship Is What Scholars Do

It is appropriate at this juncture to point out that in the past several years there has been a movement within American higher education to provide for the more complete evaluation of academics in terms of what they do. The major thrust for this has come out of a concern for the defense of excellence in teaching and the concern that academics who committed themselves to major instructional duties were not appropriately rewarded. The movement in defense of teaching and for a broader look at what academic scholars do has been led by the Carnegie Foundation for the Advancement of Teaching. In his 1990 book, *Scholarship Reconsidered*, Ernest L. Boyer, president of the Carnegie Foundation for the Advancement of Teaching from 1979 until 1995, writes, "We believe the time has come to move beyond the tired old 'teaching versus research' debate and give the familiar and honorable term 'scholarship' a broader, more capacious meaning, one that brings legitimacy to the full scope of academic work" (Boyer 1990, 16).

Boyer goes on to discuss the *scholarship of discovery*, the *scholarship of integration,* the *scholarship of application,* and the *scholarship of teaching.* The Boyer book and these definitions of scholarship have joined and helped to shape a national debate about the work of university faculty. In a follow-up to *Scholarship Reconsidered,* the Carnegie Foundation has issued *Scholarship Assessed,* by Glassick, Huber, and Maeroff (1997) to continue to influence that national debate and to provide guidance to the academic community and individual institutions wishing to consider changing the basis of faculty evaluations. "The challenge to the academic community was and continues to be the need to expand the definition of legitimate faculty work in ways that put research in proper perspective without doing it harm" (Glassick, Huber and Maeroff 1997, 11).

Glassick, Huber, and Maeroff report considerable progress and debate within academic communities to define scholarship more broadly and to more effectively measure and reward what it is that academic scholars do. Their evidence is expressed in terms of reports by university provosts about campus discussions aimed at redefining scholarship. However, there is only one major research institution with which this writer is familiar that has actually changed the definitions of scholarly activity such that there is a formal recognition in the evaluation of faculty of the different kinds of things they do.

Oregon State University is the first Carnegie I Research university to make significant changes in the definitions and evaluation of scholarship on a university-wide basis. In the early 1990s Dr. Conrad "Bud" Weiser was dean of the College of Agriculture at Oregon State University. His efforts to introduce a broader view of scholarship led ultimately to a university-wide initiative finally resulting in a change in the definitions under which faculty would be promoted and tenured. In 1995, in a unanimous decision by its faculty senate, Oregon State University adopted the following definition of scholarship for use in the evaluation of faculty:

> Scholarship is original intellectual work which is communicated and the significance is validated by peers. Scholarship may emerge from teaching, research or other responsibilities. Scholarship may take many forms including, but not limited to: research contributing to a body of knowledge; development of new technology, materials or methods; integration of knowledge or technology leading to new interpretations or applications; creation and interpretation in the arts (OSU Faculty Handbook 1999).

According to Weiser, the Carnegie definition of scholarship and that used at Oregon State University were developed to accomplish similar objectives. Both articulate more comprehensive visions of scholarship that can be used as a basis for recognizing, evaluating, and rewarding faculty across all university disciplines and missions. However, he argues, the Oregon State University model is broader and suggests that scholarship is not the exclusive domain of academia. "These models and others yet evolving," he writes, "will hopefully accelerate progress towards making the criteria and processes used to evaluate and reward the faculty more congruent with the missions of universities" (Weiser 1997).

Changing the coin of the realm is a very important and daunting task. The Carnegie Foundation has contributed significantly to a national debate, particularly on behalf of the teaching mission of universities. The Oregon State University effort led by Weiser, engaging all of the faculty of that institution, is also significant because the Oregon State definitions of the realm are broader than the Carnegie definitions, and because at OSU they have carried them through to implementation across an entire research university. At the end of the 20th century, these are promising signs of change in the view of what scholarship is. However, the prevailing norms of scholarship are still those dominated by the definitions of excellence as set forth by the scholarly communities (professional associations). Those communities place their emphasis on peer-reviewed publication in the journals they control and attend very little to the other missions of the university.

Conclusion

Given the structure of incentives that dominate the academy, it is not surprising that there is a desire among administrators of public universities to find ways to increase public service output. There is widespread recognition by administrators that there is much on campus that could be translated into direct public service, if the right stimulus could be found—there are excess capacities to be exploited and spillovers to be generated. As public funds decline, the need for evidence of direct public service is urgent for university leaders seeking to maintain budgets from the public sector.

Because public service activities by academics are still substantially outside of the realm of scholarly evaluation, the practice of many scholars to consult for fees, and indeed to set up private businesses that use the university as "mail drops," may be viewed as a logical consequence of the incentive system described here, and of the limited ability of administrators to deal with such issues within the policies of the university.

Notwithstanding the benefits to scholars and scholarship from consulting, it is not public service. It provides relatively little political payoff to the university, particularly when compared to direct public service activities. Indeed, in many circumstances universities hire professional, nonacademics, who are for the most part beyond the control of the scholarly communities, to organize and perform direct public service programs. This is even true within the extension service programs in colleges of agriculture, despite the long tradition of public service and formal assignments for their faculty. In some colleges of agriculture, totally separate extension departments have been organized in order to protect faculty with primarily extension assignments from the inappropriate application of the evaluative standards of the disciplinary communities. Given the discussed view of the important contributions that engagement can contribute to scholarship, this is really a bizarre, and self-defeating institutional arrangement.

The character of the political economy of the academic community described here is such that the problem of eliciting public service activity from academics by university leaders is a formidable one. It is the contention of this analysis that unless the existing incentive system is taken into account in the design of programs to promote public service, the chance for success is meager at best.

There is a special calling—a public commission—for scholars and for the universities that support and sustain them. Since the Morrill Act and the establishment of the land-grant universities, there is a special commission for public universities to serve in direct ways the public whose taxes provide for them. Given the character of the community of scholars there is considerable difficulty in getting public universities, and the scholars they sustain, to live up to their public commission. The challenge of statesmanship in the modern university is not so much to defend and protect the allegiance of scholars to scholarship, but rather to obtain the allegiance of scholars and their communities to the interests of society at large. Both the universities' and the societies' survival may depend on it.

4

From Theory to Practice in the Agricultural Sciences

Introduction

In the preceding two chapters we have discussed issues of the character and sources of public service, outreach, engagement, and extension in public universities, specifically the land-grant universities. We have examined the possible contributions of engagement to the relevance of science and to the practice of the scholarship of discovery. We have examined the culture and norms of the academic community. The ways that academics are rewarded and the control of the reward system, we have argued, mitigates against faculty members becoming interested or involved in outreach or extension, despite a strong interest on the part of university administrators to have an active public service, outreach and extension program.

In this chapter, we move from generalities that cross many disciplines, most universities, and generic public service, to an example that is specific to a single discipline, and to faculty with extension appointments within the land-grant universities. In some degree, this is a move from theory to practice, though we believe the example is yet another test of the workability of the notions being here set forth. In some parts of the chapter, the first-person voice is used because the arguments are based on the author's personal experiences.[1]

Throughout land-grant universities' histories there have been implicit, and sometimes explicit, differences in practice, policies, and self-image—culture—between different segments of the scientific community. At Virginia Tech since the early 1990s, the partition is between the "229 budget colleges" and the rest of the campus. The 229 budget colleges are the College of Agriculture and Life Sciences, the College of Human Resources and Education, the College of Forestry and Wildlife Resources, and the College of Veterinary Medicine, hereafter called "core land-grant colleges." Each of these units at Virginia Tech receives federal formula funds administered by the U.S. Department of Agriculture (USDA) and state matching funds, as well as state direct and supplemental appropriations, all in support of both research and extension programs. The same is true for core land-grant colleges at land-grant universities throughout the country. The agenda of spending these monies is in those areas generally agreed upon by the federal congress in its approval of the program of the USDA. Budget category 229 is the Virginia Tech budget category that accounts for the state supplemental monies for these core land-grant colleges' research and extension activities.

It is in these core land-grant colleges throughout the country, whatever their names, that there is the greatest appreciation for applied problem-solving research and for extension/outreach, even within the land-grant universities. However, as indicated in earlier chapters, even among scholars who examine the productivity of the agricultural science system, most of which embodies the same core land-grant colleges, there is an ignoring or deprecation by omission of the extension/outreach function.

Just because there is greater appreciation for outreach and applied problem solving within the community of the core land-grant colleges, does not mean that it is regarded equally. Greater appreciation may only mean that there is a larger number and proportion of faculty members with outreach obligations than in other colleges of the university, and they persist in speaking their mind on the subject within their departments and within their disciplines. Further, the images within academic departments of land-grant universities about what it is that extension people do and how they do it are confused and unclear. "You mean they give you an office and tell you to do good things?" This confusion is true even within those departments that have some history with faculty engaged full- or part-time in extension, and it is certainly the case in the departments of the university without an extension tradition. The details of the work that faculty members with major extension appointments do are not well understood.

Why Academics in Extension Are Not at the Cutting Edge

Extension faculty members spend ungodly amounts of time on the telephone and on the road. They seem to have a penchant for collecting all matter of reference materials such that many have incredibly cluttered offices. If they are in the office on any given day and participating in the coffee room conversations of the profession, many are disdainful of the discussions of the nuances of various methodological issues that seem a large part of the concern of many researchers. Indeed, some with heavy extension assignments are down right hostile to or disparaging of disciplinary research and researchers, both of which they describe as esoteric. They are scornful of the researchers' seeming inability and/or lack of interest in addressing the complexities of the kinds of problems the extension specialist is facing.

That kind of behavior, which appears hostile to some of the cutting edge issues of the profession and its advocacy for practical knowledge, contributes to the confusion about what extension faculty do and think. It leads faculty without extension appointments to frequently associate those in extension with a lack of concern for "excellence" or "rigor" in scholarship. It is certainly clear to their detractors that extension faculty are not at the cutting edge of their profession. "What, in God's name, are such 'antiacademic' people doing in academic departments?" That confusion becomes particularly explicit and is associated with considerable conflict when the system attempts to evaluate the performance of faculty members with major extension assignments.

Part of the confusion that clouds the discussion and generates the conflict comes about because of a lack of a shared understanding by many people within the system, both those with principally research assignments as well as those with principally extension assignments, about the institutional history and character of the land-grant system itself. However, simply understanding more about the system is not likely to evoke much in the way of a reduction in the conflict since the dissonance is the result of more than just ignorance of history. The conflict is the result of institutionalized sets of incentives that evoke conflicting behavior and intellectual priorities from faculty with primarily extension appointments as compared to those with teaching and research responsibility.

Given the character of the disparate institutional incentives that operate on researchers and extension faculty respectively, there exists an opportunity, perhaps even an obligation, for extension faculty to move more clearly to the disciplinary cutting edge in part by more aggressively participating in defining the edge. In order to develop this reasoning, it will be necessary to describe in greater detail than in the preceding chapter the incentives to which researchers respond, as well as a discussion of who defines the edge. It will also be important to describe the incentives to which those with major extension appointments are responsive. Finally, there will be an attempt to set forth a strategy for extension faculty that will help them have their interests prevail in the conflict over who will define what is the intellectual cutting edge of the discipline. While the arguments made here may be generalized to any academic department within a land-grant university, the comments are based principally on the author's personal experiences and observations of several agricultural economics departments.

Because the land-grant universities were crafted over the years by the political actions of individuals who came off farms and became middle class professionals with a mission they wanted accomplished—the application of science to rural problems—mechanisms were established to see that the scholars stayed on course. They politicized and democratized the scholarly agenda. Thus the application of science to agriculture was both a scholarly act and a political one and each was democratizing. Many of the established control mechanisms have been eliminated or rendered ineffective by a variety of changes in the university, in agriculture, and in the society.

The land-grant model was designed principally as a means to keeping the academic scientists' feet to the fire with respect to the type of research that was to be accomplished. It is this writer's view that this was the fundamental character of the system and not its administrative combining of teaching, research, and extension as often was cited. Crafting the arrangements to control the scientists required political support; maintaining it also requires support. In addition to the nonformal education role that extension was established to play in response to the political demands of agricultural interests, it is also a major political arm of the system, collecting grass roots support from the clients it serves.

The following restatement of the circumstances necessary for an extension program to be able to earn and collect credit from clientele is helpful in understanding the principles underlying the land-grant model, and conditions that are necessary to its successful performance:

- Complex organizations in the public sector, like colleges or universities, require sustained political support for their continued financing and vitality. In exchange for support, the institution's program is influenced by its clients.
- The application of science to somebody's problem is a political act and has the potential to generate political capital on behalf of the institution that produces the scientific solution.
- The magnitude of the political capital generated from new knowledge is a function of the number of users, of its net value to each, and the political and economic clout of those users.
- The collection of political capital usually requires a separate transaction from the capital-generating act of introducing users to the innovation. The solicitation and collection of political support frequently involves either assisting clients to organize interest groups or establishing relationships with existing interest groups. Such is the history of the relationship between extension and organizations like the American Farm Bureau, which at the county level was the early expression of farmers' interest in having a county agent.

- Sustained political support for a publicly funded organization like a land-grant university requires that it produce a sustained flow of useful information to its clients. The institutionalized test of the usefulness (relevance) of the research agenda that produces the information is in part a function of beneficiaries' willingness to act on behalf of the provider of the information. Providing audiences an opportunity to be heard about the scholarly agenda reduces their costs in articulating their views. If that process of testing the relevance of the scholarship is effective, it will influence what scholarship is at the "cutting edge," in part by assisting in sustaining budgetary support for the work.
- The engagement described by the preceding conditions will modify the scholarly agenda of the university (college/department/scholar) making it more relevant—more demand than supply driven—and can improve the quality of the research.

This model is readily comprehended from the land-grant history related earlier. In several ways, the original Morrill-Wade Act of 1862 failed to accomplish its stated purpose. After students in the new land-grant colleges had studied the few standard works on scientific farming, little was available to them. It became apparent that the application of science to agriculture needed more than classrooms and students; it required new knowledge. This could only be supplied by research and experimentation (Rainsford 1972). Thus the Hatch Act of 1887 established agricultural experiment stations as an integral part of the system in each state. Still the benefits desired by agricultural interests were not forthcoming. According to Rainsford, most of the students in the land-grant colleges did not study agriculture, even though they came from farm families; results of research and instruction did not reach farmers because they stayed on the farm. In 1914, further corrective legislation in the form of the Smith-Lever Act established a Cooperative Extension Service in each state. Finally, after 52 years (1862–1914), their purpose appeared to be achievable, and in the ensuing 50 years or so, the land-grant system appears to have stayed the course and achieved its purpose.

This brief recounting of the history describes the mechanisms that enabled the production and distribution of the products of land-grant science. In the early days and for about 50 years following the Smith-Lever Act, "enablement" also meant control of the research agenda and implicitly control of the definition of the "cutting edge" in the several disciplines of agricultural science.

Within universities throughout the world, few other communities of scholars, if any, were publicly funded on such a continuing basis.

However, those who paid the piper expected to, did, and do, call the tune. Indeed, Mayer and Mayer (1974) described the agricultural science establishment—the land-grant colleges of agriculture and the USDA—as an "island empire." On one hand, they speak with considerable admiration of the high degree of productivity and systematic approach to mission-oriented research. They then proceed to disparage the system as second-class science because it is separate from the rest of science. There is reason to believe they meant that agricultural science suffered from separation from the science in the private universities, in part presumably because Tufts University, where they were at the time of their writing, was not within the circle of universities included in the agricultural science establishment.

The Changed Political Economy of Scholarship in Colleges of Agriculture: Who Controls the Research Agenda?

In one of the most widely circulated discussions of changes within the land-grant universities, Schuh (1986) identifies the attitudes of scholars as a prominent symptom of "malaise" within the system today. It is, he asserts, the "pervasive attitude . . . that applied work is not important; publishing for professional peers and consulting for the highest paying firm or government agency are the priority tasks" (Schuh 1986, 6). Schuh makes clear that commensurate with this distortion in priorities, a sustained flow of information benefits directed to nonstudent clients of the university is not forthcoming.

Some insight into influences on the behavior and the attitudinal norms within the scientific community in colleges of agriculture is shown by the research of Busch and Lacy (1983). Table 4.1 is from that work and lists criteria that influence the choice of research problems among agricultural scientists. Some criteria are internal and essentially personal to the scientist while others are external. Of the external criteria, some are clearly the domain of administrative discretion or are an institutional feedback from clients. However, most of the external criteria are influences from the larger community of scholars whose acknowledgement is important to scholarly reputation and fame. Others, like the "likelihood of clear empirical results (#8)," are instrumental to fulfilling the scholarly reputation criteria.

Of the few criteria that can be controlled administratively, the "availability of research facilities" is third and "funding" is ninth. Clearly the dominant influences on the researchers' agenda in the 1980s became internalized within the value structure of the scientist or are controlled by the larger disciplinary community. This evidence clearly indicates

Table 4.1 Rank order of criteria for research problem choice among agricultural scientists

Rank	Criteria
1	Enjoy doing this kind of research
2	Importance to society (scientist's own judgement)[a]
3	Availability of research facilities
4	Scientific curiosity
5	Potential creation of new methods, useful materials/devices
6	Publication probability in professional journals
7	Client needs as assessed by you
8	Likelihood of clear empirical results
9	Funding
10	Evaluation of research by scientist in your field
11	Priorities of the research organization (college or USDA)
12	Potential contribution to scientific theory
13	Demands raised by clientele
14	Credibility of investigators doing similar work
15	Currently a "hot" topic
16	Length of time required to complete the research
17	Potential marketability of the final product
18	Colleagues' approval
19	Publication probability in experiment station bulletins/reports
20	Feedback from extension personnel
21	Publication probability in farm and/or industry journals

From: Busch and Lacy. 1983. *Science, Agriculture, and the Politics of Research.* Boulder, Colorado. Westview Press. Chapter 2, Table 2.1, pg. 45.
[a]Parenthetic comment added from interpretation of the text.

that, at the end of the 20th century in the agricultural sciences, the test of the appropriateness of scholarship and its relevance are left primarily to the scientist, to the norms he has internalized, and particularly to the controls exerted by the disciplinary communities. The Busch and Lacy research makes clear that scientists within the agricultural science establishment are committed to excellence and to work of use to society. However, scientists insist on determining for themselves what is excellent and what is useful. It is also clear that they place an enormous value on the written and published word since that is the major means they have for gaining approval from the scholarly community for those personal judgments. It is possible that the approval may be as myopic as the judgment about the scholarship in the first place.

In the past, when a larger portion of the funding came to colleges of agriculture via formula funds the test of relevance was administered by university leadership—by deans and presidents. It was fundamentally a political budgetary test. Most research resources, whether from state funds or federal formula funds, came through the college or university and were administered by that leadership. The scholarship that was supported was that which passed the test of relevance as seen by university and client group leaders. More recently, farming practices sometimes puts one group of farmers in conflict with another, or puts farmers in conflict with others in society, for example, environmentalists. In these settings, the test of relevance in the hands of university administrators is much less valid, and administrators run the risk of appearing like advocates for one side or the other.

There clearly has been a change in the relative power of administrators vis-a-vis faculty within the land-grant universities in the last 40 years. Further, the shift in the test of the relevance of scholarship away from the influence of interest groups acting through university administrators and state budgets appears commensurate with the decline in the relative power of university administrators. Indeed, Schuh suggests a causal relationship between the two when he identifies the strengthening of the authority of university administrators as a major means to ending the "malaise" of the land-grant university. Unfortunately his urging for greater authority to administrators does not suggest how that can be done and assumes simplistically that the decline in authority is the cause of the malaise.

Certainly there is little to be gained by the hiring of tyrants as administrators and there is ample evidence and many stories about the tyrannical rule of deans and presidents of land-grant universities when they had greater authority. Further, it would be an error to think that the only way to regain an institutionalized test of scholarly relevance that truly reflects university constituents is to return authoritarian rule and all the vagaries that go with it. Increasingly, university administrators are organizing formal advisory groups to assist in priority setting for all aspects of university activity. However, as will be discussed in the case of integrated pest management programs, the advice and political support that is received depends on who is brought to the table.

This analysis leads us to examine the reasons for the changes in the relative power of the various actors within the political economy of agricultural science at the land-grant universities since, either directly or indirectly, they are the reasons for the change in the test of scholarly relevance. These same factors have influenced the norms that land-grant agricultural scientists have internalized and are perpetuated by the scholarly disciplines—the dominant criteria influencing the choice of research problem.

Funding Changes in Support of Colleges of Agriculture

The decline in formula funds to colleges of agriculture, in both relative and real terms, and a commensurate increase in competitive grant and contract funds going directly to faculty members has made increasing numbers of faculty more independent of university administrators for their scholarship support. It is useful to remember that the funding important in influencing the relative power of the actors on campus is not the great bulk of the budget that goes for faculty salaries or operation of the physical plant. It is rather the small amounts of money that are available to support research assistantships, supplies, travel, and other items that are operational or support funding and are usually discretionary in someone's budget.

Clearly there are many issues relating to funding of land-grant science and the behavior that is elicited under different arrangements well beyond the scope of this book. In terms of this analysis, the mainline agricultural experiment station budgets buy science resources and outcomes that are intended to have public good dimensions. The grant and contract funds from the private sector produce results that may have public good dimensions but are more likely to also have private good attributes that can be captured and exploited by the private sector. That is indeed the motive for the grant or contract with the university in the first place.

In general, economists argue, where the benefits of funding have public good attributes, there is a high likelihood of the public underinvesting in the activity. Where funding produces benefits that can be captured by private firms, there are strong incentives to invest until the returns on the investment are fully exploited. Since turning down grant and contract research is virtually unknown, the recovery and distribution of "overhead costs" on university grants and contracts with the private sector completely misses the point with respect to the impact of such funding on the research agenda and the political economy of the university.

The Changing Power of Agricultural Constituents in the Society

The decline of the relative power of agricultural interests within the society has made it more difficult for those interests and the deans they supported to act unilaterally without regard to others who would make claims on the system. When the Morrill Act was passed in 1862 farm people accounted for about 50 percent of the U.S. population; at the turn of the 21st century they account for less than 2 percent (Drabenstott 1999). As other publics such as environmentalists make claims on the system, agricultural interests appeared to be able to retain control of the research agenda, beyond their ability to support it. The result has been that colleges of agriculture have had relatively little scientific output to offer new constituencies and have therefore had difficulty in gaining much support from nonagricultural clients.

Further, just as there are market failures in product markets because of attributes of the goods, there are difficulties in political markets associated with the characteristics of the client group and the information they are provided (McDowell 1985). Unlike farmers, most nonfarming clients are not tied to a specific location. Even the reason they seek extension information may be related to a role they play beyond their job, such as being a locally elected official or a parent. Collecting support from such clients over time is more difficult than collecting from farmers because they are also more mobile than farmers—a condition that must be accommodated when serving them.

The role of agricultural interests in the control of both the extension agenda and the research agenda in colleges of agriculture are one of the most significant problems and opportunities to be dealt with as the land-grant universities move into the 21st century.

The Changing Size, Program and Governance of the University

Much of the growth within the land-grant universities since the end of World War II, influenced initially by the GI Bill, has been in the enrollment and scope of the residential instruction program. Much of that enrollment took place outside the colleges of agriculture and the other land-grant core colleges, and the rest of the university community tended to reflect negatively on the "cow college" image as being unscholarly—certainly not the image that was wanted. Commensurate with the growth in students and faculty outside the core land-grant colleges was a tendency to look to the private colleges and universities, particularly to the schools in the image of the European institutions, for the images of what a university should be. This was in part because of the increasing influence of faculty with scholarly origins from those institutions.

Additional changes accompanied the growth of the land-grant universities and changed the internal power relationships. The numbers and size of disciplinary departments as separate entities rose and faculty governance procedures were strengthened. The latter reduced the authority of administrators particularly when that authority was unpopular or appeared capricious. The growth of the disciplinary departments further strengthened the view of the sanctity of disciplinary research as compared to problem-solving or multidisciplinary research, both of which are more directly related to client service and applied problem solving.

National Science Policy

The funding of national science policy after WWII focused on disciplinary or basic research through the establishment of the National Science Foundation and the program of the National Institutes of Health. It was largely directed to non-land-grant universities according to Bonnen

(1986). Because that policy and its funding were principally directed to disciplinary scholarship it had a major influence on the prevailing norms and views about scholarship—the values, beliefs, and politics—within the major scientific and disciplinary associations and on the campuses.

According to Bonnen, because much of the scholarship in colleges of agriculture was problem solving, it has suffered doubly from the prevailing science policy. Under the attitudes promoted by that national science policy, much of the scholarship of agricultural colleges was not considered appropriate for funding, nor was the scholarship that was accomplished by them considered particularly prestigious. Indeed, because of the separation of the land-grant agricultural science from mainstream science, agricultural science was, and continues to be, mostly ignored in discussions of national science policy. Even at the level of the study of science and technology as practiced in the Center for Science and Technology Studies at Virginia Tech there is virtually no attention given to the study of agricultural science.

The major conclusion of this section is that for a variety of reasons, some of which we have attempted to detail, the leadership of the land-grant universities no longer control or much influence the research or scholarly agenda in the universities. Indeed, it appears that they only really did so when agricultural science was the dominant science practiced. The professors are in control (or out of control, depending on your point of view), and the definition of the "cutting edge" of science is left almost entirely to them.

Agricultural Economics and Agricultural Economics Extension—A Case Study

Those faculty members with major extension assignments in the core land-grant colleges understand their responsibility in terms of providing a variety of deliverable information services to the audience(s) associated with their commodity, sector, or subject matter area. In pursuit of those objectives, one is relatively free to determine just how or what he wishes to do so long as there is evidence that something is being done. Extension Specialists (the language internal to cooperative extension in most states) in agricultural economics spend considerable amounts of time in the following activities:

- attempting to identify and understand the clients' problems;
- understanding the setting or context of the problems—the institutions, identity of important political actors, and the technical dimensions; and
- developing or obtaining information that will contribute to solving the problem.

The first two account for part of the large amounts of time spent on the phone, and the last point explains that many extension economists keep lots of the materials that cross their desk since it might be useful at a future date when someone asks a question. Most of this activity is covered under the "production and delivery of a sustained flow of information to clients" part of the land-grant model.

Because their assignments give them principally a client/problem orientation, extension economists of necessity evaluate the materials they come across in terms of usefulness to those problems or clients. In that regard, most find little of use in the *American Journal of Agricultural Economics* or in other greater or lesser scholarly economic journals. Some of the trade or lobby group publications are useful in keeping track of the political-institutional environment, and some of the practitioner society magazines are useful in problem definition, identification, and even in suggesting approaches to solutions.

To its credit, the American Agricultural Economics Association has published a popular, policy-oriented publication, *Choices,* since 1986, which many find helpful in staying abreast of some of the policy issues. In 1997, the AAEA established yet another journal, *Review of Agricultural Economics,* which is intended to be less methodological and more catholic with respect to publishing the work that is what Agricultural Economists actually do. Bonnen (1986) points out that the profession is schizophrenic with respect to its journals, a diagnosis that is even manifest in this book that liberally cites the *American Journal of Agricultural Economics.*

For the most part, extension economists are not at the "cutting edge" of the profession because, wherever it is, it is not where they are; and where the edge is, is substantially irrelevant to what they are trying to do. However, by their own evaluations, little of the material that would be useful to extension clients and in extension programs would be publishable in any of the scholarly journals. At the same time, virtually all in extension are acutely aware of the great need, individually and collectively, for help in solving the great array of problems that extension clients face. Many extension economists know, too, that addressing many of the problems they encounter may well require very complex methodologies, many of which are beyond the ken of most in extension.

Many in economics extension are dismayed because the opportunity cost of much of the effort by research colleagues is very high and appears to have neither short-term nor long-term relevance. They are further distressed by the guise that a particular piece of research is on an "applied problem" but the research that is carried out never will, nor can be, applied to the problem it ostensibly purports to solve.

Many extension economists are not at the "edge" and question the relevance of those at the "edge." They are where they are because they are paid to be there, not because they are better, or worse, than those at the edge. Unfortunately, where the "cutting edge" is in agricultural economics may be where it is substantially because of the failure of extension economists to behave as scholars.

In Search of the Edge

Unfortunately, finding the "cutting edge" in agricultural economics can be more easily accomplished by describing where it *isn't* rather than where it *is*. Schuh (1986) says of scholarship in the land-grant system generally (and presumably as an agricultural economist he includes his own discipline in his generalization) that it is uninterested in applied practical work. In a rather remarkable trilogy of papers from the 1984 American Agricultural Economics Association meeting in Ithaca, New York, Swanson (1984), Hoch (1984), and Barkley (1984) take a hard look at "the mainstream" in agricultural economics and conclude that not only is it uninterested in applied practical work, it may no longer be capable of doing it. Swanson (1984) quoted Beneke as follows:

> There exists throughout the profession differences among its members as to what constitutes appropriate priorities for professional effort and accomplishment. One dichotomy . . . involves tool making, extending, modifying, and refining on the one hand and tool using, or problem solving, on the other. . . . It seems to me that one group is saying that what really counts is the person's command of theory and his (her) capacity to develop and improve upon modern analytical tools. I do not hear the opposite argument, that the real test of competence is the capacity to use analytical tools effectively. There also seems to be a few among us arguing that problem solving research is a worthwhile activity only if the tools used in the process are erudite. I rarely hear concern that the problem studied may be a trivial one (Beneke 1983).

Barkley (1984) identifies the ability to recognize a significant problem as perhaps the most neglected part of the training (and presumably practice) of contemporary agricultural economists. "Our skill (in the application of quantitative methods) has torn us from the problems we seek to solve." In that regard, he identifies three areas of thought where the ability of agricultural economists to identify problems, form hypotheses, gather data, test, refute, and eventually provide answers to local problems and instruct policy formation has come into question. The arenas he identifies are farms, institutional economics, and income distribution.

Barkley (1984) further argues that one of the most basic lessons in the discipline of the last 75 years is that much real problem solving involves crucial variables that are nonquantifiable. "That they are unmeasurable does not mean that they are unponderable" (Barkley 1984, 801). The revival of rationalism—the exercise of reason—as a supplement to empiricism, he argues, will strengthen our profession in three major ways. "First, our professional discourse will include think pieces as well as quantitative pieces. Second, we will apply our best tools in their most favorable light. Third, we will bring new and exciting perspectives to the solution of problems and to the formation of policies for the rural United States" (Barkley 1984, 801).

Hoch, in his piece, "Retooling the Mainstream," says that he is asking the question, "Are we up the creek (mainstream) with an ornate paddle? (Hock 1984, 793)" He appears to conclude that we are.

There is considerable evidence that wherever the "cutting edge" is in agricultural economics, it is substantially irrelevant to solving practical problems. Further, suggests Bonnen (1986), since something less than 10 percent of the material that is published in the *AJAE* would meet the disciplinary standards of the journals in the economics profession, most of what agricultural economists call "high-quality" research does not qualify as either problem-solving or disciplinary research. If that is also true in other disciplines in the land-grant system, then the land-grant science system itself, is in jeopardy, since as has been argued, the successful functioning of the land-grant model of science requires an institutional test of objectivity and of relevance, which are provided by engagement.

Extension Economists' Responsibility—An Indictment and an Agenda

A summary of the previous section identifies the mislocation of the "cutting edge" in agricultural economics in the following dimensions:

- The profession and practice of agricultural economics does not identify and rationally describe real problems, and sees no value in doing so.
- Too many of the people familiar with the empirical tools are unfamiliar with what is most important to apply them to.
- There is too little criticism of work within agricultural economics simply on the basis that the work is trivial.
- The profession has ignored important areas in which agricultural economics has been successful in the past, but with which it no longer appears able to deal. Specifically identified were farms, institutional economics, and income distribution.

The character of the assignment of agricultural economists as extension specialists, as described above, would appear to make them the members of the profession with the greatest comparative advantage to speak to precisely these issues about the mislocation of the cutting edge, and yet they have not spoken out. When the rules affecting the determination of what would be considered good scholarship changed, extension faculty never figured out the new ball game or even the playing field. As a matter of fact, many continue to hold either the administrators or individual researchers responsible for the dysfunction in the discipline when, individually and collectively, extension economists are as responsible as either other group.

Not only have many extension economists failed to write in a scholarly fashion for their peers, many do not much use the written word as the basis of their own extension programs within the state. Because it is possible to carry out the appearance of the extension function without the written word does not mean it is an appropriate way to do it. It is this writer's suspicion that there are several reasons that so much extension gets done by extension specialists in all fields without written materials:

- Time saving—if you can get away with winging it, why not?
- Self-preservation—when information is particularized to a user via a personal consultative type of relationship, the first and primary source to which the information is attributed by the user is to the person of the extension specialist, not the institution he represents. Extension specialists use that proclivity by clients associated with personalized distribution of extension information to build direct personal political support.
- Avoidance of scrutiny—if you don't write it down, it is a lot easier to get away with fuzzy economics, biology, or engineering, actual misinformation, and/or undefended opinion.
- Frustration—if you can't get scholarly credit for it anyway, why bother.

The failure to write is simply a fundamental violation of the obligations of being "delegated intellects" (Taylor 1981, 23) protected by tenure and the principle of academic freedom. Further, it is politically an error. If the slogan for the researcher is "publish or perish" then for the extension scholar it is, "publish or you have no program." There are those who would argue that written materials are not sufficient to the establishment of an extension program and with that this writer agrees. However, we argue that written materials are a necessary condition to a program that distributes new knowledge, makes it possible to collect support from the beneficiaries of the information, and facilitates evaluation of the scholarly content of the program. It is hard to know who is worse for the land-grant system, the researcher who does irrelevant research or the extension specialist who has no program.

It has been acknowledged already that the writing necessary and important for extension purposes is quite different than what is currently (or maybe ever) appropriate to the journals. However, the process of developing and preparing written materials for an extension program requires a brand of scholarship already described as decidedly lacking and needed by the agricultural economics profession. The transference of that kind of scholarship into publishable think pieces, conceptual descriptions of the policy arena or its institutional setting, critiques of the trivia being published, or good problem descriptions should be relatively easy.

Those of us with extension appointments are among the only folks who can move the "edge" in a direction that will result in greater scholarship that is relevant to the problem solving we claim to speak for. Given the character of the land-grant model, the salvation of the system may be up to those scholars with major extension assignments—but they (we) must behave as scholars.

Conclusion

There are four main observations or conclusions to be made from the discussion of this chapter, much of which is explicitly about agricultural economists with appointments obligating them to spend a substantial amount of their time working on behalf of programs delivered in the name of cooperative extension.

First, the discussion explicitly argues that it is most likely these scholars—the extension specialists—that can contribute to some of the most pressing conceptual and relevance issues in the discipline.

Second, the argument about the disciplinary contribution of the extension economists illustrates the benefits of "engagement" asserted by the Kellogg Commission publication *Returning to Our Roots: The Engaged Institution* (Kellogg Commission 1998) and by the formal arguments of Chapter 3.

Third, there is indeed a difference between the academic cultures in the core land-grant colleges and the rest of the colleges in the land-grant universities. The future of the land-grant universities into the 21st century will very likely depend ever less on the programs and engagement of those core colleges and ever more on the programs and engagement elsewhere in the universities—in business colleges, colleges of engineering, medical schools, colleges of architecture and planning, colleges of education, as well as the colleges of arts and sciences. The problem of establishing outreach/extension programs that systematically and formally engage those faculty members with the society is formidable indeed. As the case example makes clear, if difficulties within the favorable environment of the college of agriculture already lead to dysfunctional behavior by extension faculty, how will it be possible to elicit the necessary engagement in the rest of the university.

Finally, the assertions of this chapter vis-a-vis the evaluation of extension work and extension specialists further illustrate the importance to extension and outreach programming of making changes in the definitions of scholarship and the evaluation of academics.

Whether or not cooperative extension can broaden its funding and program portfolio and take the lead in the institutional changes necessary to promote that larger engagement is uncertain.

Notes

1. Much of the thinking for this chapter was influenced by a paper prepared by the author and presented at a workshop for extension agricultural economists entitled Maintaining the Cutting Edge, at the American Agricultural Economics Association annual meetings in East Lansing, Michigan, July 31, 1987. The paper is: McDowell, George R., Why Many Extension Economists Are Not at the Cutting Edge and What They Can Do About Moving the Edge. Unpublished, 1987.

5

Cooperative Extension—Part of the Problem or Part of the Solution?

Introduction

In Chapter 4, the assertion was made that in order to achieve the goal of an engaged university as set forth by the Kellogg Commission it would require considerable effort, particularly for the schools and colleges of the university that have little land-grant extension heritage. However, before getting to that question, it is appropriate to ask about the current state of the Cooperative Extension System and its ability to contribute to the future engagement of land-grant universities. It is to that question that we now turn.

As indicated in Chapter 1, there is still about 15,000 full time equivalent (FTE) staff employed by Cooperative Extension programs throughout the nation. In 1991, that number was 15,876, which was down from an all-time high of 16,954 FTEs in 1983 (Ahearn 1999). Total extension expenditures in 1997 were $1.483 billion. There have been and are considerably more individuals than the number of FTEs formally associated with, and paid by extension, since many university faculty members have partial appointments with extension. Extension staff are, by and large, hard-working and creative.

There are extension offices in about 70 percent of the 3,066 counties of the nation providing nonformal, functional educational programming to all areas of the country on a wide array of subjects. In most cases, the counties without an extension office are served by staff located in adjacent counties or in a regional office. Extension actually has more field offices than the number of counties because there are offices in many of the nation's cities. Cooperative Extension has an absolutely remarkable record of performance and reputation. For many, many years, Cooperative Extension was considered the most reliable and unbiased source of information for rural people, particularly for farming people (Feller, Kaltreider, Madden, Moore, and Sims 1984). That reputation still persists and is, in this writer's view, well earned. That is, the information provided by Cooperative Extension is well grounded in research, and the organization serves well those people that it serves.

Chapters 3 and 4 presented considerable discussion about the distortions in the university incentive systems that give rise to dysfunctional behavior by university campus faculty, at least as relates to their valuing and performing public service. It is not that they are bad people; it is just that their behavior as elicited by the culture of the academy is not supportive of extension and the original concept of the land-grant university. This chapter is an attempt to describe some of the distortions within the extension system that give rise to dysfunction within it, and, thus, that will influence its ability to contribute to the land-grant universities engaging American society in the 21st century. Again, the issue is not a question of good or bad people, but of the incentives that may or may not give rise to behavior that will serve the needs of the university into the 21st century.

Part of this chapter examines the question of whom Cooperative Extension, as one of the outreach arms of "the people's university," actually serves. In the context of this discussion, we will talk of the portfolio of extension programs and extension audiences as being an indication of breadth or lack of breadth of the program and the people served. Much of the dysfunction within extension, it will be argued, is within the agricultural programs and by staff who serve farm audiences.

In 1991, the following assertions were made by this writer to the leaders of extension agricultural programs from all across the county (McDowell 1991, 1–2):

> I propose to argue that the USDA/land-grant extension system has been captured by agricultural audiences. It has been taken hostage to such an extent that it can no longer function effectively to inform agricultural audiences of some of the most important issues facing them. Further, I will argue, because it has been taken hostage, it cannot expand its scope or fulfill its mission with respect to nonfarming audiences.
>
> When taken together with the other problems in the university, extension, held hostage, will not make the 20 more years that some of its most severe critics give it. Unless there is profound change, I believe it will be virtually dead in 10 years. My friend, Tim Wallace, from California, says he is already in mourning.
>
> There are two other introductory comments that need be made before we get started. First, those of us in the USDA/land-grant system who argue for significant change, particularly for a broader agenda—for research and extension programs that reach new clients—are not antiagriculture, as some think we are. As a matter of fact, most of us believe that the future ability of the land-grant system to serve agriculture and farming audiences depends on the support of nonfarming people.

Second, it is important that I acknowledge that many of you are striving, with great effort, to figure a way out of the consequences of the dysfunction I will attempt to describe.

By the time of publication of this book, the 10 years since the 1991 prediction should be about up. In 1999 at the time of this writing, much of the dysfunction within extension continues, but the tenacity and survivability of the extension system is greater than this writer ever estimated.

So, aside from this author's assertions, how does the portfolio of extension sort out in terms of the program areas of agriculture, home economics, 4-H, and community resource development? Table 5.1 provides that information for selected years over the 20 years from 1973 to 1992. More recent data is unavailable. Table 5.1 indicates that the 2 percent of the American people who farmed in 1992 still got the lion's share of extension resources, and not a lot has changed since then at least by casual observation. Further, as is plainly evident, agricultural programs gained at the expense of 4-H and community resource development programs.

It appears that the decline in the numbers of farmers (14 percent between 1975 and 1987), the declining fortunes of rural people and places, and the plight of rural people generally have resulted in no change in the portfolio of extension programming, at least as measured by these broad administrative categories. Rather, it seems evident that, as their fortunes declined, farm interests have increased their grip on, and dominance of, the extension agenda.

Table 5.1 National, state, and local extension professional FTEs by program areas, 1973, 1982, 1987, and 1992.[a]

	Percent of Professional FTEs			
	1973	1982	1987	1992
Agriculture and natural resources	38	44	46	47
Home economics	21	22	23	24
4-H and youth development	32	27	25	22
Community and rural development	9	7	6	7

Source: PDE-ES-USDA, May 11, 1992.
[a] More recent data was sought from CSREES/USDA to bring this table up to the late 1990s. The USDA was unable to provide the data by these or any other categories that would permit an estimation of these categories and still sum to 100 percent of the FTEs.

All of this is true at the same time that many of the private institutions serving farming have made significant changes in their portfolios. For example the Springfield, Massachusetts, and the Baltimore, Maryland, Farm Credit Banks sued the farm credit system in about 1985 for raiding funds from them to help bail out the rest of the system that was in financial trouble. The two East Coast banks were not in as bad shape as the rest of the system during the 1980s, in part because they had a broader portfolio that included many more different kinds of enterprises than the portfolios of the other banks (Swackhamer 1999). Similarly, many of the farm implement dealers, and other input suppliers that have survived, have done so by appealing in their product mix to many nonfarming customers. In the case of farm implement dealers, the broadening has been toward both construction equipment and homeowner lawn and garden equipment.

There is little evidence of response from the extension/land-grant system to the larger issues in rural America—the hostage taking is almost complete. In 1992, the last year for which there is data, the extent of the nation-wide commitment of extension to rural development (Community Resource Development) was 7 percent, as measured by staff full time equivalents (FTEs) and nothing has much changed in the intervening years. However, there is a caveat that must be acknowledged before making that as a final conclusion. The caveat follows from the respective roles and relationship between the extension and the research function in the system.

While it is inappropriate to view extension simply as the pipe through which knowledge generated by research is disseminated, it is no coincidence that with respect to existing extension programs, the area of greatest research investment and support is also the area of greatest extension commitment of resources. To switch to a computer system analogy, as a long time extension economist, I do not view extension field staff as "dumb terminals." There is, nevertheless, considerably more for extension front line people to work with where there is greater research support to their programs. Further, as set forth earlier, the value of information to extension clients, not in aggregate but individually, has a great deal to do with the political economy of the system, and is directly related to the character of the research being carried out.

In this context then it is very difficult for extension to move into areas ahead of any, or much, research support to its programs. Thus, more important than who controls and establishes the extension agenda is who controls the research agenda. The previous several chapters have focused a great deal on those issues.

With extension captured by agricultural interests, and with the research agenda and the university captured by the professors, it is not surprising that there is little evidence of a broadening of the extension program or clientele portfolio. But how has the suborning or hostage-taking of extension taken place?

Contemporary Extension

In light of the historical development of extension both politically and institutionally as detailed in earlier chapters, the purposes of the Cooperative Extension Service are, and have always been, the following:

- To seek to know the problems of ordinary people and to bring those problems to the attention of the researchers.
- To deliver functional education, based on the best scholarship available, to ordinary people, to help solve their problems.
- To collect political support from the beneficiaries of extension programs in order to fund the continued research and education of ordinary people of the society—not just, or even primarily, farmers.

University faculty are protected by tenure in order that their science will be objectively carried out and be the best science possible. It also obliges faculty to speak, because they can speak with impunity, of the implications of their science to the society, particularly if those implications are unpopular or not widely understood.

The extension system and its staff are similarly protected, some by university tenure and faculty status and all by a unique institutional setting. That protection should enable extension field staff to influence the research agenda with an identification and description of problems based on independence of thought—not based on short run political gain. That freedom should also enable extension field staff to act as educators, because they can speak with impunity about the issues that affect the groups with whom they work, even if the messages they have to tell are unpopular. Because extension staff are protected, they are obliged to speak, particularly to the unpopular but important issues.

It is in precisely this nexus—this link between being the objective educator with respect to unpopular messages and the institutional maintenance job of collecting support for the system—that has caused the Cooperative Extension Service to have enough dysfunction to be taken hostage by the farm groups.

The USDA/Cooperative Extension System, as part of the land-grant universities, is one of the special political institutions of our society. It has contributed significantly to the continued productivity of American agriculture. In part because of extension, agriculture remains one of a few sectors in which the United States continues to have an international comparative advantage. We can all be proud of that. It is also important that we continue to invest in, and demand, excellence in research and extension programs that serve farmers—commercial farmers in particular.

The extension system's ability to endure confounds both those who admire it, and those who would do away with it. It clearly challenges this author's 1991 prediction about the system's demise. There is, in extension's three-tier financing and control, an incredible ability to be independent, to be shielded from short run political manipulation from what ever level it emerges, to deal with controversy and survive. There is no other agency or institution of our society that is so constructed and so protected. Extension financing in 1997 was a combination of federal, state, and local (county) with about 24 percent federal, 49 percent state, and 21 percent local with the balance of 6 percent from grants and contracts, on average nationally. In the university, the professors are individually protected. But neither the university, nor the colleges of agriculture, apart from extension, is as well-protected as is extension with its grass-roots connections.

As a result of that protection, extension people have been able to speak freely to policy and practices of federal, state, and local government. For the most part, they are not apologists for national farm policy. Indeed, they have many times been critics of it. The down side at the federal level was reported by Dr. William Wood, Extension Specialist, University of California, Riverside, when he led the 1978–79 evaluation of extension. Wood (1978–79) became aware of the exasperation of the Office of Management and Budget (OMB) at the thought that federally-appropriated funds would be put into a program that the federal government couldn't control, and into programs that were not totally supportive, and even sometimes critical, of federal policies.

Extension people are not apologists for the state departments of agriculture. In fact, in many states extension staff are in conflict with them, from time to time—partly because extension is not subordinate to them, as some departments of agriculture would like, partly because Cooperative Extension gets state funding for "agriculture," which does not come to the state agency, and partly because extension is frequently not in agreement with their programs or approaches to the agricultural issues of the state.

The state level down side is reported by Sandra Batie, Professor of Agricultural Economics, at Michigan State University (Batie 1991). Dr. Batie spent a sabbatical year at the National Governors Association and tells of hearing midwestern governors complain that extension was beyond their control. They put money into extension but couldn't direct it. More than one told Dr. Batie they were going to do whatever they could to cut it off.

County agents have greater freedom of action than do other county workers or employees. Even in the states where they are county employees, they have special freedoms. When they run afoul of county politics, unless they have clearly misbehaved, they will more likely be transferred than fired.

Another down side is that, as the number of farmers and their influences has been decreasing at county, state, and national levels, there is less and less support at the grass-roots level for agricultural programs. In response to this, Virginia and several other states have in the past five years or more, been going to a cluster form of field staff organization where some agricultural agents are shared with multiple counties—and that's not because anyone wants it that way.

This rather elaborate three-tiered institutionalized protection is substantial. It is analogous to tenure for the professors, and some county staff have university tenure as well. But how often do you hear a professor speak of tenure as an obligation, as opposed to a job perk? Like the professors, extension administrators and staff have not always used wisely the freedom that they have. Some have even abused it. Few deans, directors of extension, or agricultural program leaders have understood that it was their obligation to lead and educate farm groups about what was in their best interests, whether they wanted to hear it or not. More often the extension system and the people in it follow the farm groups around like bulls following cows in heat.

The major result of dysfunction in the extension system is the degree to which the Cooperative Extension Service has been captured by farming interests. Extension staff are not apologists for national farm policy—that is true. Extension staff are not apologists for state or local policies and programs—that is also true. But many of extension's agricultural staff—administrators, specialists, and agents—are apologists for farmers and ranchers and are therefore no longer objective educators.

The March 20, 1991, *Chronicle of Higher Education* published an article suggesting that a major part of the problem of the land-grant agricultural science system is that it has been captured by the agricultural chemical companies (Jaschik 1991). While there is a serious problem in the control of the research agenda, neither the research agenda nor the influence of the chemical companies can explain the anger and outrage that can be evoked from many county agricultural agents at the suggestion that organic sustainable production methods may be as valid or more valid than traditional chemical approaches.

This writer suggests that the reason for the emotional response to direct or indirect criticism of contemporary production methods is because of the excessive identification of agricultural agents with their farmer clients. They have, in the language of the Peace Corps and international work, "gone native." Even the suggestion that there might be a more socially acceptable way for farmers to farm is condemned as scientific heresy—regardless of the scientific evidence. This may explain as much of the land-grant image as lackey for the chemical companies as does the chemical companies' actual influence on the research agenda. If that is not loss of objectivity, what is?

How the Extension Service Became a Hostage of Agriculture

There are a number of ways that the hostage taking of extension has happened and some of those continue to be management problems for extension administrators. As indicated earlier, the loss of objectivity and hostage taking turns on the dilemma of the necessity of collecting support for the system and being educator as well. To review—the conditions that are necessary in order for an extension program to be able to earn and collect credit from clientele are the following:

- Positive Net Benefit Condition—The program must generate a positive net benefit—the total benefits of the education or information must be more than what it costs to get it, including time and travel.
- Attribution Condition—Most of the net benefits, regardless of magnitude, must be attributed to extension.
- The Solicitation Condition—The collection of political capital usually involves a separate transaction. The clients must be identifiable and thus susceptible to being solicited for support.
- The Political Action Condition—Acting politically for extension must cost the clients less than their past and anticipated future benefits. As with all agencies in the public sector, extension does a variety of things to reduce the costs of political action including taking constituents to Washington, D.C., to meet with congressman. (McDowell 1985)

Extension works hard to design programs that meet all of the conditions set forth above. To fail to do so would be foolish. One of the most effective ways to design programs to satisfy these conditions is to particularize information so that it is, or appears to be, absolutely specific to an individual client. Efforts so designed have the advantage of most easily meeting the attribution and solicitation conditions. Soil tests and computer-generated information programs have this advantage especially when the information is requested by clients. Farm groups meet the solicitation condition relatively easily because you always know where to find them—they have one foot tied to the ground—when you want to collect political support from them.

It is in the details of meeting these necessary conditions that extension staff and leadership have lost their way and have been taken hostage by farm and commodity groups. Again, as with the professors, the behavior is not because they are bad people. On the contrary, they are hard working and dedicated. They unfortunately do things that are dysfunctional to the long-term survival of the institution. Consider the following:

The Personalizing of Extension Contacts with Farm Audiences

In the politics of universities—of tenure and of extension specialists and agents perceiving themselves on the short end of the stick—some extension specialists have made use of these notions to their own advantage, generating support for themselves rather than for the institution.

Computers are one way of particularizing information. A program like FINPAC (financial management package) will do a farm financial analysis based on the specific information for a particular farm—it is unique to the particular farm and of little use to a different farmer. Another way to particularize information to a specific client is to personalize the delivery of extension information, like a consultant, to individual producers, or even a group of producers. The information is then mostly attributed to the individual agent or extension specialist rather than to the extension organization. In order to solicit support for the organization, extension leadership must go to the specialist or the agent to gather the group's support.

This tendency to stand between the organization and the clients is part of the reason that some extension specialists don't much build their programs based on written materials, as was argued in the preceding chapter. If they present their information in written form—with the extension indicia and all of that—then the institution can claim a larger share of the credit—the attribution condition. Further, if the information is written and published, the specialist's presence is not always necessary for conveyance of the information.

Once the dependency relationship for political support is established between specialists and extension, specialists use that support and power for their own support within the organization. The problem is that if the political capital accumulated by agents and specialists is used by them to shift college and extension resources from cattle to grain programs, from 4-H to agriculture programs, from community development to range management programs, then you can't use it to increase the total pot. Once the specialist or agent has sold his soul, the extension organization gets taken hostage with him.

Those state extension services that place great emphasis on extension specialists traveling (chasing hood ornaments) to hold the hands of the client groups and less time in scholarship, have bought into the hostage taking.

Reactive versus Proactive Programming

Another way that the hostage taking has occurred is similar to the discussion above about the behavior of specialists, and for the very same reasons. In this case, it specifically involves the behavior of field staff. When agents spend their time and energies in a way that conveys that they are at the beck and call of farmers and have no independent programs of their own, the message is very clear. Even worse, if your only program is to wait around until farmers call, or you have so much time that you can afford to help run all the local cattle sales, then the hostage taking is already accomplished at the local level.

It is clear that some level of personalized contact and reactive program delivery is important to stay abreast of farmers' problems, and to be viewed by them as being aware of those problems and thus a credible source of information for farmers. But if there is not an aggressive proactive program, based on independent thought and empirical evidence, about the problems of farmers in the state, then it is not possible to provide any leadership to farmers. It is certainly not possible to tell them some of the uncomfortable truths about a changing agriculture.

It is this writer's observation that there is more reactive programming in the agriculture program than there is in any of the other programs. The client numbers are so enormous, or the problems so different, that extension staff in home economics and 4-H have not gotten into this form of dysfunction very much. Individual, on demand, service programming, as a means to particularizing information, is simply not a feasible way to reach very large audiences. Indeed, in the early years of extension, it was not a very feasible approach for the agricultural program because of the large number of farmers and the limited staff. It is true that throughout extension's history many county agents have had favorites among their clients and spent disproportionate amounts of time with them.

In Virginia, there was a conflict between field staff in farm management and some pre-tenure state specialists who were trying to implement a statewide proactive program. The field staff argued that the Management, Analysis, and Planning Program (MAP) they were implementing in a proactive way was okay, but what they really needed was the specialists, on demand as consultants. Where there is an understanding of extension and a willingness to evaluate it, there must be evidence of scholarship and the "consulting" mode of delivery simply does not produce evidence of scholarship that can be used in a tenure portfolio. Further, the number of farms a consultant can reach is many fewer than can be served by a well-designed proactive program that meets a real need that farmers have.

One of the leadership group in Virginia's Cooperative Extension Service reports that in 1999, the agricultural extension program in Virginia was virtually a consulting service to farmers and that providing agricultural agents with in-service training for carrying out of educational programs was a waste of time for both instructors and agents. It was reported that when agricultural agents were asked to describe their program day, the reply was that on any given day, they did not know what they were going to do until they got their phone messages—that is strictly reactive programming.

Wasting of Political Capital

The arguments being made here are that the political process that is necessary to maintain the support base for the extension system is best understood as an exchange or *quid pro quo* system. The land-grant university produces new knowledge and disseminates it. In exchange, clients are asked to provide support via the public budgetary process to enable the institution to continue. When the system produces knowledge that is of interest to more than one constituency and does not collect from all, it is wasting its political capital.

A classic case of this type of dysfunction is with respect to integrated pest management programs. The IPM money was originally delivered to the land-grant/extension system somewhere around 1973 by environmental interests who wanted to reduce pesticide use. There has been some absolutely fantastic science produced and some very significant gains made in the reduction of pesticide usage. But there has been, to this writer's knowledge, no concerted effort at either the state or the federal level to formally report to environmental groups about the pesticide reduction accomplishments of the program. Even more critical, there has been little or no effort to formally build the environmental groups into any kind of a council or advisory group with respect to pest management research or extension.

In 1991, Dr. Ann Sorensen, an entomologist hired by the National Farm Bureau Federation, organized a national IPM coalition with Farm Bureau's very nervous support. It was only 18 years late—and only lasted for a couple of years because Dr. Sorensen left the Farm Bureau. At the state level, there have been 25 years of political capital that extension, the agricultural establishment, and the farm groups have failed to collect on. Further, they've had a chance to wear environmental "white hats" and have blown it!

If that's not dysfunction, what is?

Control of the Extension Agenda

Farming people make up about 2 percent of the population. It's that way in Virginia, and it's just about that way nationally. No matter how you cut it, the budget for extension has been, and is, disproportionately committed to farmers and ranchers. Depending on whether you call it "agricultural competitiveness and profitability" as one of seven national base programs, or "agriculture" as over against 4-H youth, home economics, natural resources, and community resource development, it comprises about 40 percent of the system's resources. The total is about 15,000 FTE professionals or about $1.2 billion annually. A substantial part of the rest of the budget also goes to farmers and their families in the form of 4-H and family programs. Even without making the earlier point about extension and land-grant universities not being especially for farmers, the above allocation between programs is generous, if not excessive, on behalf of agriculture.

Despite this, farm groups, whose ability to deliver extension budgets has declined at least in proportion to their numbers, constantly seek to control ever more tightly the ever-declining pot. The *Hoard's Dairyman* editorial on February 10, 1991, is a fine example. "Agricultural Extension Is Under Attack," they said and reported that "a handful of federal extension administrators are determined to get extension out of agriculture and into 'societal issues'"(Hoard's Dairyman 1991). They urged dairy farmers to exert their influence on the internal allocation of extension resources through their county extension offices, the state extension offices, and even federal congressional offices. To emphasize their point they published figures on the numbers of state dairy specialists, by states, in 1980 and 1990. The total decline they reported was 33 percent.

Dr. Patrick Boyle, then extension chancellor of the University of Wisconsin, wrote to the dairy magazine to correct some of their misstatements and errors and challenged their numbers. He assured them of extension's continued support to dairy programs, as right he should. What he didn't point out was that there was a 37 percent decline in the number of dairy farms in the country between 1978 and 1987 (a close approximation of the same period). Even if Hoard's numbers were correct, which Boyle said were not, it can be argued that dairy farmers had a 4 percent gain in extension resources. At the same time, extension was only committing 5 percent of total resources to natural resource programming and 6 percent to community resource development.

In the control of the extension agenda on behalf of farm and agricultural interests, the deans of the colleges of agriculture are frequently the point persons on the campus for the farm and agricultural interest groups. If, as with agents and specialists, their relationship with the farm groups is reactive rather than proactive, agricultural interests will act as though the dean is in their pocket, whether he or she wants to be there or not.

Control of the Institutional Setting of Cooperative Extension within the University

Some control of the extension agenda and of the hostage taking by agricultural interests gets played out in land-grant universities as forces within and without the university recognize the enormous success of the Cooperative Extension Service on behalf of farmers and ask "why not have extension for everyone?" That recognition has led many extension directors and university administrators to seek to do one of two things: either elevate the leadership of Cooperative Extension to a post that has university-wide responsibilities, or make the leadership of extension responsible to such a university-wide post.

In many states, the dean of the college of agriculture is, or was, "dean and director" indicating that he was dean of the College of Agriculture, director of the Agricultural Experiment Station, and director of the Cooperative Extension Service. At Pennsylvania State University, the separation of the leadership of extension from the dean of the College of Agriculture only took place in 1996–97. The arrangement was so common in the system that it became a part of the lore of the brilliance of the land-grant design. It was, in the view of many deans, what made the land-grant system unique. Some land-grant leaders even understood that having the administrative control of research, teaching and extension under a single individual could lead to differences in research relevance and quality. Few of those deans of Agriculture, it seems, ever wondered whether these advantages could be achieved on behalf of other applications of science and other sectors of the economy, or by other administrative and leadership means.

Until the movement to broaden the portfolio of extension started in the mid-1960s, Cooperative Extension was entirely under the control of the leadership of the core land-grant college deans, usually the dean of the college of agriculture. Some of the most ardent and passionate internecine battles within land-grant universities in the past 40 years have taken place over maintaining or recovering control of the Cooperative

Extension system by colleges of agriculture. The struggles over Cooperative Extension at Virginia Tech are a classic case and are by no means unique.

By establishing an Extension Division in 1967, Virginia Polytechnic Institute and State University was one of the earliest of the land-grant universities to broaden the vision of extension to a university-wide mission under leadership with university-wide authority. John Dooley, currently Virginia Tech's associate extension director for 4-H and Family and Consumer Science, and assistant dean, College of Human Resources and Education, tells the story of Virginia Tech in his 1998 Ph.D. dissertation (Dooley 1998).

According to Dooley, the primary architects of broadening the Virginia Tech extension program to a university-wide effort in the mid-1960s were the university president, T. Marshall Hahn, and extension director, William Skelton. Hahn, who served as president of Virginia Tech from 1962 until 1974, had a vision to transform a predominantly agricultural and engineering institution into a major comprehensive university that would serve all of the people of the commonwealth. Skelton, who served as state 4-H Leader from 1950 to 1962, extension director from 1962 until 1967, and dean of the Extension Division from 1967 until 1976, believed the extension approach that was so productive for farming people could be used to assist all of the citizens of the commonwealth. Together Hahn and Skelton orchestrated the passage of state legislation that provided for and formalized a funding structure for all three missions of the university. "Instead of Virginia Tech being one funded agency of the commonwealth, it would become three: Virginia Tech Division of Instruction, Virginia Tech Division of Research, and Virginia Tech Division of Extension" (Dooley 1998, 89).

According to Dooley (1998) in considering the approach they would take, Hahn and Skelton considered two basic organizational structures. The first was a "College of Extension and Continuing Education" under whose auspices all extension and continuing education would be organized into a single college. The second model was more audacious by calling for a restructuring of the entire university. All off-campus special educational activities and all programs directed to nontraditional students would be under the authority of a single administrative officer, the dean of Extension. However, all faculty would retain their academic appointments in their respective academic units. Faculty within all units of the university would and could have different assignments in teaching, research, and extension. When Hahn and Skelton chose the second approach, it was the land-grant model, so long a part of the colleges of agriculture, applied to the whole university. Virginia Tech was on the way to be what the Kellogg Commission described in 1998 as an "engaged university."

As Hahn, Skelton, and other Virginia Tech leaders implemented this effort in the late 1960s, Dooley (1998) reports they struggled with whether the new division would be called "public service," "outreach," or "extension," and all were being considered. Skelton recalled the discussions in an interview with Dooley:

> Other universities in the state carried out public service. Only Virginia Tech and Virginia State were legally authorized to do extension. The people already knew and could identify with the term extension. We (Hahn, Brandt, and Skelton) spent hours upon hours debating this issue. We conscientiously made the decision that whichever model we pursued, its trademark would be "extension" (Dooley 1998, 90).

According to Dooley (1998), in the late 1960s and 1970s the Extension Division became the leading edge for a growing and dynamic university and it was a very good time for Virginia Tech. Hahn, says Dooley, attributed much of the university's success to the establishment of the Extension Division.

> The strengthened Extension Division, with a program presence across all colleges of the university, provided us a vehicle through which we could reach all areas of the state with our programs. It was this enhanced capacity to meet the needs of all the people that gained us political favor and allowed for our unusual growth in all areas of the university. The greater the ties that we had to the people of the state through extension, the greater our success was with all our initiatives on behalf of the university (Dooley 1998, 99).

But all was not well among the agricultural interests of the state of Virginia who took a jaundiced look at the ever-increasing resources going into the enlarged and broadened extension system. They appear to have had the view that all those resources could have been going into agricultural programs. Dooley (1998) quotes President Lavery, who succeeded Hahn, as saying that "something happened in Halifax County" in 1978 that encouraged state representative Frank Slayton to challenge the agenda of the Extension Division and to argue that what it was doing was not commensurate with its land-grant origins and history. The result was that over the next several years, Virginia state government and the Virginia legislature undertook to critique every aspect of the Extension Division program. By 1980, the Joint Legislative Audit and Review Commission (JLARC), the investigative arm of the state Legislature, had made recommendations and demanded a mission statement that would delimit the Extension Division's programs. That started the Extension Division on a long but continuous slide and a return of Cooperative Extension to being under the control of the College of Agriculture and Life Sciences.

Extension's return to the college of agriculture nest was finally accomplished in 1989 (22 years after establishment of the Extension Division at the university level) when President James McComas divided the already reduced Extension Division into two parts, Cooperative Extension and the Division of Continuing Education. He moved Cooperative Extension back into the College of Agriculture and Life Sciences.

McComas came to Virginia Tech with the reported intention of further expanding and enhancing the outreach function of the university. But he was put on notice by the Virginia Farm Bureau and other agricultural interests who visited him in Ohio while he was still President Designate, before he ever got to the Blacksburg, Virginia, campus. The dean of agriculture brokered the visit during which McComas was informed that if he wanted any support from the agricultural community for anything, he would have to return Cooperative Extension to the College of Agriculture.

When legislation was enacted in 1994 reversing the 1967 legislation that had established the three funding streams for the university, and the extension budget was folded back under the funding for the Agricultural Experiment Station, the death knell for a broader extension role in university outreach was sounded at Virginia Tech. What had happened in Halifax County in 1978 was that some farm clients who wanted an extension program hostage to them raised questions with sympathetic county commissioners and thence to their delegate in the state legislature. With limited vision and a self-serving view of history they set about to "fix" extension—for themselves.

The final insult to any vision of a broader extension portfolio for Virginia Cooperative Extension occurred in 1995 when the Virginia General Assembly included new language in the appropriations act for extension. The appropriation act for 1995 stated, "It is the intent of the General Assembly that the Cooperative Extension Service gives highest priority to programs which comprised the original mission of the Extension Service, especially agriculture at the local level" (Commonwealth of Virginia 1995, 158). The purpose was explicitly to limit the community resource development (CRD) program, a federally mandated part of the extension portfolio, among others.

It is widely understood that the CRD program was explicitly identified in language contained in the governor's confidential working papers, but that language never made it into the final appropriations bill. The common knowledge within Virginia extension about the source of the restriction in the appropriations language and the impetus for its inclusion is that senior agricultural extension agents with personal relationships with powerful state legislators suggested the language be included. Dean of Agriculture, L.A. Swiger, confirmed the much-rumored information to be true by stating in response to the direct question, that the agricultural agents thought that by cutting off the right leg, the left leg would grow bigger (Swiger 1998).

There are similar tales to be told in West Virginia, Michigan, and Minnesota, among other places. Similarly, the promise of change taking place at Pennsylvania State University, Iowa State University, and Oregon State University among others in the late 1990s seemed much less permanent and much more fragile in light of the 30-year history in Virginia.

That farm interests in Virginia wish to own, if they do not already own, the Virginia Cooperative Extension Service and the College of Agriculture and Life Sciences is evident in a 1998 exchange between staff of the Virginia Farm Bureau and extension faculty and leadership of the Department of Agricultural and Applied Economics. At issue was a report of research on nonpoint pollution and the economic and environmental impacts of nutrient loss reductions on dairy and dairy/poultry farms that was due to be published for distribution in a publication series of the department.

When the Virginia Farm Bureau first heard about the manuscript they asked for an opportunity to read it before it was published. In the course of providing that courtesy, the senior lobbyist of the organization suggested that the Farm Bureau should be given the opportunity to review, and implicitly approve, all policy-oriented extension publications that the department produces. The lobbyist was told that was not possible. However, no one in a leadership role in the college took the opportunity to explain to Farm Bureau leadership why even the suggestion of editorial review was inappropriate, or further, why it was not likely even in the self-interest of the Farm Bureau.

Because of the obvious sensitivity to the publication, the authors undertook a major round of reviews and reworking of their results to be sure they were on sound footing. Then in order to assure that everyone within the university or outside the university was given a chance to be informed of the results and comment, the authors held seminars on campus and off to assure all that the sky was not falling. This whole process delayed the release of the publication six months longer than would normally be the case.

Conclusion

The Cooperative Extension system nationwide is a substantial organization with considerable record of success, and a great deal of goodwill and trust from its several audiences and the general public. However, as we have argued in this chapter, there are reasons for concern about extension's ability to contribute to the future of the land-grant universities.

The general thrust of this chapter has argued that Cooperative Extension can be, and in some places has been, taken hostage by agricultural interests to such an extent that it cannot even serve well the educational needs of agricultural audiences. The latter is the subject matter of Chapter 6. We have described the mechanisms and behavior entrapments that can and do lead to individual and institutional compromise of Cooperative Extension, restricting the organization's portfolio and allowing it to be dominated by agricultural programs. In many places this substantially limits extension serving as a broad-based outreach arm of land-grant universities.

However, before writing off Cooperative Extension's possibility and potential to play a major role in the engagement of land-grant universities with American society into the 21st century, it will be necessary to examine further both the problems and the promise of the extension system. Further, extension varies a great deal from state to state and from university to university. More examination of that variation needs also to be accomplished before rendering any final verdict on the future of extension in the land-grant universities into the 21st century. It is to these questions that we now turn.

6

Agricultural Extension—How Well Do the Hostages Serve the Hostage Takers?

Introduction

In the previous chapter, we asserted and attempted to give evidence that the Cooperative Extension Service in many states and counties has been captured and held hostage by agricultural interests. Some of the mechanisms whereby the hostage-taking occurs were described. It was also asserted that in many places Cooperative Extension and the affiliated agricultural research agenda have been taken hostage to such an extent that it can no longer function effectively to inform agricultural audiences of some of the most important issues facing them. This chapter addresses that issue.

There is reason for the reader to ask, "Why spend time delving into the agricultural extension program when the broader issue this book addresses is about the future of the land-grant university and the extension system within that university?" Even at this point it is apparent that a central thesis of the book is how to use the successful experience of the agricultural extension model on behalf of other audiences, rather than recounting the past or present accomplishments of the agricultural extension programs. However, the ways in which the most successful and most clearly supported programs of extension serve their clients, in the face of the problems facing those clients, speaks to the future potential of the organization to act as an agent of engagement of the entire university. Further, because there are massive changes taking place within the relationship between the society and its public universities including the land-grant universities, and within the extension services around the country, it is appropriate to examine the ways in which the system is responding to its dominant audience, farmers.

Indeed, it is within the portion of the extension program serving farming and the agricultural economy that the change and stress are most evident because of the declining size of that audience and the pressures to change the way agricultural extension programs are managed. Thus, the commitment of resources to agricultural programs, the historical and thus symbolic importance of agricultural programs to the extension model, and the changes taking place in the circumstances of American agriculture require some examination of agricultural extension programs.

Forces of Change in American Agriculture

The following are a number of the most significant forces effecting change in the agricultural sector of the American economy.[1]

- Agriculture and farmers will hence forth be viewed as "on-the-dole" and policies and appropriations on their behalf will no longer be sacred cows.

  The budgetary excesses of the farm programs (as much as $50 billion in 1986) of the 1985 Farm Bill in part led to the 1995 Federal Agricultural Improvement Reform (FAIR) Act of 1995 and its open market and market stimulating approaches. Notwithstanding the attempt of the FAIR Act to change the on-the-dole image of American farmers, the $6–8 billion emergency aid to farmers in 1998, along with the fiscal excesses of the past, are part of the reason for changing attitudes about government support of farmers and their programs.

- Farming will continue to be closely scrutinized on environmental grounds, and farmers will consistently lose to environmentalists if they force a choice.

  The demand for farm commodities is more income inelastic than is the demand for environmental quality. As our economy improves and incomes grow people will want more environmental quality, but the amount of food they want will not grow proportionally.

- Agricultural and other food products are increasingly being linked to commodity attributes at the production level.

  For example, certain types of soybeans are more suitable for soy sauce than the varieties most frequently grown. In order to gain the premium price for beans in the soy sauce market, farmers must grow the right beans. Thus it is that many more farm product markets are beginning to look like markets for differentiated products. These changes have a great deal to say about both the markets and the management of the firms within the markets. It may very well be that Demming's total quality management (TQM) will be the new farm management and marketing of the future.

- Disintermediation—the reduction in the numbers of transactions and actors in the production, processing, and distribution of products from production to retailing.

Disintermediation is increasingly the rule in many agricultural commodity markets. Some of this is accomplished by the greater influence of vertical integration and/or vertical coordination through contracts and other coordination mechanisms. Though not unrelated to the greater differentiation in agricultural commodities markets, there appear to be other forces that are influencing that change, not the least of which is the inadequacy of government maintained commodity grades and standards to reflect consumer preferences. Many consumers consider a "choice" cut of beef too fat, and most "choice" steaks are too tough by any stretch of the imagination.

The use of universal product codes (bar codes) at retail establishments and self-identifying discount cards that permit tracking of associated purchases by individuals permits reduction of inventories and just-in-time supply. Greater amounts of information about purchasing patterns and practices give retailers greater power in the chain and further reduces transaction costs.

- The greater influence of international markets and the internationalization of domestic markets have resulted in an increasingly complex set of finance and marketing circumstances for many agricultural commodities.
- The proposition of Castle (1989), that farming may be an industry with a constantly declining or flat long-run average cost curve, appears to be increasingly in evidence.

That is, economies of size appear to continue to hammer smaller producers with the result that fewer and fewer producers are producing the major portion of output. When coupled with the previous five propositions, agriculture is increasingly in the hands of the strong, and the strong are increasingly big.

- Risk, and the varying perceptions and valuation of risk by consumers in the marketplace and with respect to the environment, are having a profound effect on both food production and processing.

Included in this issue of consumer risk are the greater awareness, and faddism, in dietary knowledge and its impact on consumer demand. There is greater consumer scrutiny, and a commensurate increase in demand for food and fiber attributes that sometimes only include methods of production, e.g., organically produced vegetables or animal products from humanely treated animals, independent of known evidence of product differences or effects.

Some producer groups have been slow to react to these signals from the marketplace. These shifts in demand, and the mysteries of consumer behavior, are likely to increasingly affect producers and will likely continue to baffle, confound, and elude many producer groups for some time to come, though not the Archer-Daniels-Midlands "supermarket to the world," or the Smithfield/Carroll Foods of the pork industry. These large corporate producers and processors are very responsive to consumer behavior (see discussion of bar codes in item on disintermediation *above*) and will produce very different products from the small, competitive, and independent farmers like "the other white meat."

- The growing distrust of science in Western societies has a strong effect on agriculture, which is itself ever more dependent on science.

    The growth in productivity of American agriculture is substantially a success story of the long-term application of science to agriculture, and of the productivity of the agricultural science establishment. However, Western society generally is increasingly skeptical of all science including agricultural science. There is a growing suspicion that all of the costs associated with past scientific advances were not fully accounted for, much less paid for, and there is much less willingness to believe the claims for new scientific solutions, and advances (Bromley and McGuire 1991). Indeed, the serious side of the 1990 novel, *Jurassic Park*, by Michael Crichton, is about the abuse of high-tech biological engineering for profit, despite the public risks—science run amok. While this distrust of science is not unrelated to the risk issues discussed in the item above, it is a different dimension and relates directly to the future role of land-grant universities.

- The rural infrastructure serving both agriculture and rural communities is in disrepair.

    The current status of that infrastructure is, in many places, inadequate to serve either a modern agriculture or the present and future demands of many rural communities.

    In the case of some of the commercial and business infrastructure serving agriculture, as the number of farms producing or using a particular commodity declines, the level of production is inadequate to support viable businesses dealing in that product. In such cases, as in farm machinery, the number of dealers declines and the character of the remaining businesses change and diversify into other product lines. As with dairy in a number of places in Virginia, there are simply not enough dairy farms to support even liquid milk collection and bulking facilities.

- The character of the scientific advances being applied to agriculture is different than in the past.

    Increasingly, the value of agriculture-based products is in the value added after the farm gate. This is clearly related to the trend toward greater differentiation of agriculture-based products identified above.

It seems evident that this value-added and differentiation is based on science and technology that is either applied to commodities after the farm gate, or that influence the character of the use of the products in processing and consumption. Much of that science is driven by consumer demand in the final marketplace and much of that research is private, proprietary research. Certainly the genetics associated with the animals raised in the mega hog farms, is one such example.

While there is still production-enhancing and/or cost-of- production reducing research being carried out, much of the research directed at on-the-farm systems is aimed at assisting farmers to comply with safety or environmental requirements. Dean Swiger of Virginia Tech (1993) estimated that between one-third to one-half of the agricultural production-oriented research is so directed. While these technological changes may be fundamental to the survival of farming, many farmers view the regulations that make these practices necessary as intrusions by ignorant do-gooders who do not understand the implications of what they are requiring. They view researchers who work on problems as similarly suspect.

- Patenting of plant materials is now feasible.

With the changing ability to manipulate plant and animal genetics through biotechnology and bioengineering such as gene splicing, there is the ability to develop plant materials and seed that can be patented and otherwise made private property. This possibility changes the whole face of varietal research, moving much of it into the private sector, and implicitly reducing the public sector capacity to either review or challenge the direction that that part of agricultural science will go. This affects several of the preceding items.

- Increasingly, farmers appear to own an ever smaller proportion of the land they farm. As urbanizing pressures and other claims on land-use intrude, maintaining viable-sized farms with any kind of proximate coherence becomes an ever-increasing problem.

- The federal government presence in agriculture is changing.

The federal presence in agriculture has been both regulatory and facilitating. It appears that the "facilitative" part is falling out of favor. For example, budget cuts in the support of grades and grading, market news, and reporting of economic planning information are degrading the "public good" information that helps independent farmers survive. This is happening at the same time that the secretary of agriculture is championing and shifting funds to support a "small farmers commission." It may be that there is not much understanding of the impact of the loss of the "public goods" the USDA produces, since the impacts are very small to any individual producer. A further

example is that the Economic Research Service, at one time the largest and best social science research organization in the world, no longer has the support required to maintain that status.
- Demand for legislative solutions to the perceived ills of the large processing firms will grow.

After years of clamoring for USDA/federal agency controls over large buyers of farm products, states are starting to act themselves. For example, North Dakota has passed a law requiring all packers buying livestock in the state to fully report prices and quantities. If such legislative solutions are sporadic, and they impede private confidential pricing arrangements, it may force some relocation of processing activity in the same way that state differences in environmental legislation has influenced the location of some agricultural activity.

In an attempt to bring many of these forces of change together and examine their implications, extension farm management economist Dr. Steven C. Blank has written a somewhat controversial book, *The End of Agriculture in the American Portfolio* (1998). Under the heading "The Last Roundup," Blank says:

> American agriculture is heading for the last roundup. Our rural countryside, both the beautiful and the "visually challenged," is heading into the final stage of its economic development. As we look out across that countryside, many of us will find it impossible to imagine our country without farms and ranches. Especially at this point in our history, when American agriculture leads the world in almost every way, it is startling to think that we will not need farmers or ranchers for much longer. But it is true.
>
> To understand and appreciate the changes, we need to place farming into context. We need to strip away the romance and nostalgia surrounding agriculture and see it for what it is: a business. It is a type of business that has limited potential for long-run profits because of its competitive nature. The whole world can "do it." In America, the cost of doing it has risen to the point where it is not very profitable compared to alternative types of businesses. Thus, the people, money, and other resources invested in agriculture currently will be forced to leave for "greener pastures" (Blank 1998, 192).

It is this author's contention that if the agricultural extension program is to be judged as dealing effectively with the issues facing American agriculture, the ways these issues are addressed must be included in extension's evaluation. Of course, as has been pointed out several times already, agricultural extension programs do not act in isolation from research programs and/or the influences of farm interest groups in the several states and at the national level.

Why the Technology Transfer Extension Model Doesn't Fit Agricultural Extension So Well Anymore

The importance of the images of extension that come from the experiences of 85 years of agricultural extension cannot be ignored, particularly when there is dissonance between those images and contemporary reality in extension. Dr. Scott Peters, assistant professor of agricultural education at Cornell University, has written an enormous history of Cooperative Extension as his Ph.D. dissertation. In *Extension Work as Public Work: Reconsidering Cooperative Extension's Civic Mission* (1998) Dr. Peters argues persuasively that the earliest intent of the land-grant universities and of extension, rooted in the ideals of education of, for and by the people, was that extension work was "public work." By public work Peters means "the visible, creative efforts of a mix of people that produces things of lasting importance to our communities and society" (1998, v).

This model of extension gave way to a technology transfer model of extension that emphasized technologic change and economic efficiency in agriculture. Indeed, argues Peters, the forces for marginalizing the civic work image of extension were active even at the time of the signing of the Smith-Lever Act in 1914 and include

(1) the growth and dominance of a particular view of farming as a "business";
(2) the rise of technocratic politics;
(3) the increasing calls for cheap food to fuel urban industry; and
(4) the belief that farmers' incomes must be raised first, before any other kind of development can take place (Peters 1998, 164).

Norman Rockwell, preeminent 20th century illustrator of life in America, captured the very best of the technology-transfer image of extension in his 1948 illustration, "The Work of the County Agent," which is reproduced as a frontispiece and is the basis of the cover design.[2] A shovel full of soil is exposed from under the sod in a field not far from the barns and silo. While three generations of men from the farm family look on, the county agent tests the soil using the vials and solutions he has brought with him in his "science kit." The application of science to farming is direct, personal, and unambiguous. That the agent and his work are much respected and appreciated is clear in the faces of the farmers. This image perfectly fits the conditions necessary for an extension program to be able to earn and collect credit from the clientele it serves, as described several times in earlier chapters.

The changes that have been and are taking place in American agriculture demand extension programming with a style of educational delivery that is at odds with this ideal model as represented Norman Rockwell, notwithstanding that it was developed to improve the economic efficiency of American agriculture. Indeed, as argued earlier, that model has been highly successful as measured in terms of agricultural productivity, though perhaps not in terms of the vitality of rural communities or numbers of persons employed in farming. To understand the limitations of the technology transfer extension model to contemporary American agriculture, we need to examine 1) the changes under way in American agriculture and 2) the character of the information that is needed by farmers and the agricultural community to accommodate those changes in their strategic planning.

Increasing or maintaining profitability in farming has always been a primary objective of agricultural extension programs. Indeed, profitability in farming has even become codified into the preeminent issue for agricultural extension, under "issues programming" promulgated by the USDA. With profitability in farming as our major focus, the Virginia Tech Department of Agricultural Economics extension faculty made a presentation to the Virginia extension leadership in the early 1990s. The presentation was about the character of the agricultural extension program and the agricultural economics contribution to it. Being cognizant of the many changes taking place in agriculture, as described in the previous section, it was suggested that the following were significant influences on farm profitability in Virginia:

- the performance of national and international commodity markets;
- state, national, and international policies affecting agriculture;
- the performance of the firms and institutions serving and regulating farming and agricultural markets—both public and private, both inputs and commodities; and
- on-the-farm agricultural production technology and its management.

An examination of the extension intellectual resources at Virginia Tech committed in 1992 to agricultural programs indicates that there were more than 50 full-time equivalents (FTE) of extension faculty specialist time devoted to the on-the-farm agricultural production technology and its management. There were only six FTEs devoted to the other areas affecting farm profitability—four in the Department of Agricultural Economics and two in the Department of Food Science and Technology.

The suggestion that the intellectual resources devoted to agricultural extension were inappropriately allocated was not happily received. It would have been even less welcome to suggest that field staff resources in agricultural extension might also be similarly misdirected. Indeed, most of the field staff engaged in agricultural extension are working on issues related to on-the-farm productivity. Given the earlier comments about the style of their work as being almost entirely reactive, or essentially as consultants to individual farmers, the Rockwell illustration is a good representation.

The Virginia example of the early 1990s is emblematic of the national picture of resource commitments within agricultural extension at the close of the century. It is also consistent with the general commitment of resources to agricultural research. Consider the National Research Initiative Competitive Grants Program (NRI) for the past several years, administered by the USDA as the major source of federal nonformula funds to agricultural science.

The table below summarizes the categories of research and the funding allocations for 1992, 1998, and 1999.

Combining the 1992 NRI categories for the plant and animal systems and the natural resources and the environment, shows 85 percent of the resources on primarily production issues remarkably like the 89 percent that 50 FTEs comprise of the total 56 FTEs of Virginia Tech agricultural extension specialists in the same year. Including the natural resources category in on-the-farm technologies is defendable since much of that category is devoted to the insights necessary to adjust farming technologies to be more environmentally compatible, as pointed out earlier. The $7.6 millions for trade, markets policy, and food technology looks remarkably like the FTEs committed to before- and after-the-farm gate

Table 6.1 Research divisions and funding levels supported by the National Research Initiative (NRI), USDA

Category	FY 92 $M	%	FY 98 $M	%	FY 99 $M	%
Natural resources and the environment	17.0	18.5	16.3	18.0	19.1	17.2
Nutrition, food quality, and health	6.2	6.7	7.4	8.2	14.9	13.4
Plant systems	37.9	41.0	34.4	38.0	38.2	34.4
Animal systems	23.7	25.6	22.4	24.8	27.0	24.3
Markets, trade, and policy	3.8	4.1	3.6	4.0	4.3	3.9
New products and processes	3.8	4.1	6.3	7.0	7.6	6.8
Total	92.3	100	90.4	100	111.1	100

Source: Program Description, Guidelines for Proposal Preparation and Submission, NRICGP, CSRS, USDA, 1992 and CSREES/USDA, http://www.reeusda.gov/cgram/nri/programs/progdesc/intro.htm#FUNDING.

insights by Virginia Tech extension. The priority trend in the NRI remained essentially the same throughout the 1990s.

The National Research Initiative, it is argued, has a strong predisposition for basic, and not even applied, physical, and biological science and clearly against social and policy oriented science. That is different, it is argued, from the on-the-farm versus off-the-farm issues and dichotomy that has been identified. However, there is clearly a strong physical/biological science versus social science predisposition in the on- versus off-the-farm issues and in the allocation of the Virginia Tech or other states' extension resources.

It is now appropriate to examine the characteristics of the information, the way it is used, and the educational requirements of informing farmers and others about the off-the-farm issues affecting farm profitability. Let us examine these in light of the technology transfer model and the four necessary conditions for institutional maintenance of extension programs.

Much of the scholarship necessary to inform understanding of off-the-farm issues is in the domain of agricultural economists. Much, if not most, of the information that must be developed and provided looks much more like public policy education rather than farm management or technical agricultural information. That is, many insights needed to inform decisions by farmers and others in the food system will pertain to collective or strategic actions. Even the adoption by farmers of some of the new production technology that is directed at niche markets or other specialized markets will be more a strategic decision than a how-to production decision. Educational programming with information that informs this type of decision creates a substantially different relationship between the extension educator and client/audience than does information that, if adopted, will directly increase production output or reduce farmer costs.

By way of illustrating the notion of information for individual strategic behavior we suggest that it is important for the farm management agents in the peanut-growing area of Virginia to learn some new words and to teach those words to their farmers. One of those words is "groundnuts." What we in the United States call "peanuts" virtually everyone else in the world calls "groundnuts." When, and if, farmers learn what groundnuts are and why it is an important word to them, it will mean that they have a new awareness of the possible implications of actions by the World Trade Organization and international markets to them. Farmers may even wish to begin to make strategic adjustments in their farm businesses. Just what those adjustments might be are beyond the scope of this book, but the issue might be a reasonable topic for some scholarship, supported by a reallocation in the NRI.

Of this it is quite clear, it will require considerably greater effort and program design to ensure that an extension program about groundnuts for the peanut farmers of southeast Virginia meets the conditions for an effective extension program. Given all the discussion of the World Trade Organization in the media, and the lack of specificity with respect to the direct action that any farmer should take in response to the insight about groundnuts, few farmers will likely consider the extension program on groundnuts has provided them with a positive net benefit. Further, it is uncertain that farmers will clearly attribute their insights on the subject to extension or that they will view their appropriate but required adjustments as "positive net benefits." It is surely a more difficult situation for even Norman Rockwell to illustrate than the testing of soil fertility.

Now consider informing farmers about "disintermediation" in the food system, total quality management (TQM), and niche marketing. It is very difficult to get farm audiences, and even more difficult to get extension leadership, to recognize that working with the local baking industry or the local textile industry is indeed "wheat marketing" or "cotton marketing."

The structure of the pork industry represents one of the most recently vertically integrated subsectors in American agriculture. It is likely that internalization of transactions between processors and producers has been driven as much by the needs of processors to meet the demands by consumers for leaner meat and by the economy's attendant to processing hogs of greater uniformity of size and carcass characteristics. However, farm management economists argue that the scale economies in hog production are all captured at sizes of operations that are much smaller than the megafarms that now characterize the industry. The economists argue that the gains to genetics and other advantages gained by vertical coordination between the megafarms and processors could also have been achieved by tightly managed cooperatives of smaller hog farmers (Pease 1999)—but that never happened. Even if farmers had understood the issues, collective action is difficult to achieve and the pork processors had the action advantage. The extension education program to farmers about this opportunity and the likely action of farmers based on that education, were both nonstarters.

Consider yet another example. Because of the concerns of many nonfarmers about environmental impacts of a variety of farming practices and on-the-farm technology, even the extension programs directed toward on-the-farm technology and its management are beginning to look much like public policy education programs. There is already difficulty in generating farm audience support for agricultural public policy education when it is objective rather than advocacy for farmers.

What about farmer response to objective extension programs on environmental policy that may require changed behavior or additional investments from them? Even if they do not shoot the messengers, they are not likely to ask them to visit again.

The point is not that public policy extension programs or programs that inform farmers about the changes in the economy or society that will affect them should or should not be carried out. They should be delivered and in many places extension does carry out such programs and does them very well. The point is that because the relationship between these programs and the benefits to agricultural audiences is much less direct than most of the on-the-farm programs, there is the classical economic dilemma of investments in public goods.

In terms of the necessary conditions for generating support from extension programs, "if you don't see the benefit, don't attribute the benefits to my efforts, or both; you are not likely to think that I'm doing much for you, and won't tell anyone that I should be supported to continue doing it." If nothing else, there is less of a smile of appreciation on the face of the farmer as he learns from the extension program that he may have to invest in a manure storage tank, carry out a nutrient management plan, or consider getting out of groundnut production. That would truly challenge a Normal Rockwell to illustrate.

How Well Are Agricultural Extension Programs Serving Farmers?

In answering the question about how well the extension system is serving agricultural audiences, there are two fundamental questions to be addressed. First, "Are the farmers getting what they want from extension?" Second, "Are the farmers getting what they need from extension?"

The answer to the first question is "yes," although many will say still not enough. For a variety of reasons, some of them valid and based on experience, American farmers have a strong belief that many, if not most, of the problems they face are amenable to physical or biological science solutions—to the management of on-the-farm technology. That is what they have asked for. It is at the farm level where farmers have the greatest sense of control. Until the excesses of the 1985 farm bill, external policies and institutions were sacred cows carefully managed on farmers' behalf by their representatives, with the tacit approval of the rest of the nation. Farmers make manifest their requests for extension assistance in part through the professional cum political relationships between commodity production extension specialists and the respective commodity groups. That is also partly where the hostage taking occurs as described in Chapter 5.

Dick McGuire, at the time New York State Commissioner of Agriculture and a farmer, spoke clearly to the confidence of farmers in agricultural science on their behalf in his *CHOICES* dialogue on agricultural science and the environment with Dan Bromley in 1991. "As agricultural producers, we had seen the miracles wrought by modern chemistry and biology in the period following World War II.... Should we give up the advances of science and technology and return to a kind of eighteenth century economy? I think not" (Bromley and McGuire 1991, 6).

But are farmers getting what they need from the hostages they have taken? This writer does not think so. If there was ever an example of the inability of science and technology, specifically agricultural science, to overcome on its own, the agricultural problems of nations, one need only look at the agriculture of the former Eastern European nations, particularly Hungary. They had much very fine science directed at on-the-farm production, including Western agricultural science. That was not enough.

Clearly there are important technical and scientific dimensions of the changes taking place in American agriculture, as described earlier in this chapter. It is quite apparent that gene splicing and molecular biology will likely contribute significantly to the forces of greater product differentiation within agriculture. There will be winners and losers among farmers. There will be new niches for new products. There will be the need for farmers to understand how to position themselves strategically in the new environment, and there will be a need for educational programs about on-the-farm production technology and its management.

However, with ratios of 50 FTE extension specialists focused on on-the-farm technology and its management and six FTEs directed to off-the-farm issues, or with a National Research Initiative that has $88.5 million in science and technology and $3.8 million in markets, trade, and policy, this writer does not believe Virginia or American farmers are getting what they need. In an exchange with Steven Blank about his book, *The End of Agriculture in the American Portfolio*, and the controversy that it had and would evoke (Blank 1998), Blank replied:

> Having spent my life and career in agriculture, the story in my book is a disturbing one to me, in some ways, but I decided that the story needed telling so that it might get other people thinking and talking about the issues. I think only when the level of debate is raised to the level of the problem can that problem be addressed adequately.... As Cooperative Extension economists, we are often called upon to tell farmers and ranchers what they "need" to hear, rather than just what they "want" to hear (Blank 1998a).

Unfortunately, it appears systemic that Cooperative Extension, and the agricultural establishment that supports it around the country, is telling farmers what they want to hear rather than what they need to hear.

Conclusion

On several occasions, this book has alluded to the notion of the extension portfolio of programs and thus to the array of audiences that support the extension system and the land-grant universities that give home, care, and keeping to extension programs. It seems evident that if what has been claimed in this chapter is true—that extension-held-hostage is not serving even its primary, traditional audience well—there is not a bright future for this old and well-established program. Further, if what Blank (1998) argues in his book is true—that the end of agriculture in the American economic portfolio is at hand—it would appear to be the time to begin seriously to reassess the future of extension and its predominately agricultural programs in the land-grant university.

As a part of that assessment, it will be necessary to examine (1) the efforts undertaken to broaden the extension portfolio; (2) some systematic recounting or discussion of the opportunities missed; and (3) some of the most promising and hopeful efforts within extension that may belie the developing impression that extension is already a dinosaur. Indeed, the dinosaur conclusion about extension may be independent of whether or not the land-grant universities themselves end up as "a sort of academic Jurassic Park—of great historic interest, fascinating places to visit, but increasingly irrelevant in a world that has passed them by" (Kellogg Presidents' Commission 1996). It is unlikely that an extension service dominated by an agricultural program that gives farmers what they want, not what they need, will be capable of providing leadership to the larger university engagement into the 21st century. It is unlikely that a dinosaur will lead the way away from an academic Jurassic Park.

Notes

1. The author is indebted to his colleagues David Kenyon, Wayne Purcell, Eluned Jones, and Jim Pease for reviewing the listing of "forces of change" and commenting on them.

2. The original of this work hangs in the office of the Dean of the College of Agriculture and Life Sciences in Stockbridge Hall, University of Massachusetts, Amherst, MA.

7

How Cooperative Extension?—The County and Federal Partners

Introduction

The designation "Cooperative Extension Service," which is no longer used in a number of states but lives on in the formal name of the programs in yet other states, has its origin in, and bespeaks the tripartite support from federal, state, and county governments. By whatever title it now uses, be it "University of Missouri Outreach and Extension," "Michigan State University Extension," "Cornell Cooperative Extension," or "Texas Agricultural Extension Service," extension programs in all of the states continue to have both county and federal partners. The United States Department of Agriculture (USDA) is the federal partner, and most of the counties in the nation contribute either directly or indirectly to the funding of extension. Figure 7.1 shows the respective contributions of federal, state, local, and contract/grant (nontax) funds to extension programs across the nation for the period 1971 to 1997 in inflation-adjusted 1971 dollars. The upper line is total national extension spending in real 1971 dollars because the respective contributions are added.

In 1997, the average proportional contribution from the respective sources was 24 percent from the federal partner, 49 percent from the states, 21 percent from the counties, and 6 percent from nontax sources. At the beginning of the period, the respective proportions were federal, 40 percent; state, 39 percent; county, 19 percent; and nontax, 2 percent. The federal partner has significantly reduced its contribution and the state partner has picked up much of the slack. However, there is even greater variation in the state-by-state data for almost any single year than there is in the national data over this time period. For example, in 1997, the federal contribution in West Virginia was 58 percent and about 17 percent in New York. The decline in federal contribution in real dollars over the past 25 years is a consistent pattern across all states as well as on average.

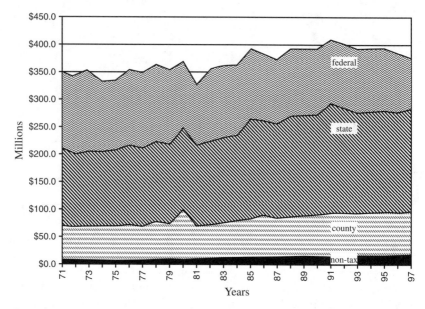

Figure 7.1 Extension expenditures by source in 1971 inflation adjusted dollars, 1971–1997.

Figure 7.2 shows the total national expenditures for extension programs in both real and nominal dollars. Total extension expenditures in 1997 were $1.485 billion. When adjusted by the Consumer Price Index, total expenditures for extension have remained virtually constant over the 26-year period, though as Figure 7.1 indicates the relative contribution of the respective partners has changed quite dramatically.

The several sources of funding available to state extension programs and the land-grant universities provide the leadership of state extension programs with considerable flexibility and independence. Even the University of Rhode Island, which receives the smallest state allocation of federal funds (not counting the amounts sent to Guam, the Virgin Islands, and other U.S. possessions), was not likely to sniff at more than $2 million from the federal government for its outreach programs in 1997. However, as will be argued in this chapter, the influence of the USDA is more significant than its proportional contribution to state extension systems. Further, the linkage between the USDA and agricultural research in the land-grant universities is yet a further influence on the extension agenda. The combined research and extension budget to the USDA's Cooperative State Research, Education, and Extension Ser-

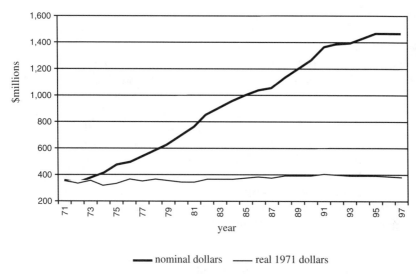

Figure 7.2 Total extension expenditures 1971–1997 in nominal and real (1971) dollars.

vice (CSREES), the agency that administers the federal research and extension monies, was $923 million in FY 1999 and $948 million proposed for FY 2000. Most of this money is passed through to the state land-grant universities, much of it on the basis of formulae. The Smith-Lever formula is based on the numbers of farms and rural residents, whereas the formula for the Expanded Food and Nutrition Education Program is based on numbers of persons living below 1.5 times the poverty line.

In the midst of the multiple bilateral relationships between the USDA and the states is the National Association of State Universities and Land-Grant Colleges (NASULGC). NASULGC serves as a coordinating organization for the interests of its land-grant membership vis-à-vis the federal government. It hosts, rather than presides over, a committee structure partly related to the flow of funds from the USDA to extension and to the Agricultural Experiment Stations. That structure is under NASULGC's Commission on Food, Environment, and Renewable Resources (CFERR) and five boards of that commission, though most knowledgeable people will agree the Board of Agriculture is dominant.

Fundamentally the elaborate committee structure of NASULGC is an effort to make the multiple bilateral relationships into a multilateral relationship where the states' interests are coordinated. Of particular importance are the ECOP (Extension Committee on Policy) and

ESCOP (Experiment Station Committee on Policy) in the overall coordination of the relationship between the states and the federal partner for extension and research issues.

Similarly, the relationships among county governments, county extension advisory councils where they exist, a state association of county committees, and the leadership of the land-grant university are some of the more complex, oft times hazardous, and sometimes rewarding aspects of the cooperative part of extension. In some states, the state association of county extension committees is one of the significant lobby efforts on behalf of both state and county extension efforts. Again, as the records show, county contributions vary widely among the several states. More important than the amount of the county contribution are both the form of the contribution to extension programs and the character of the county extension advisory board's empowerment and limitations on county extension staff activity.

In this chapter we will seek to describe, if not untangle, some of the web of influence and conflict that both confound and determine the organization, operation, portfolio, and content of the Cooperative Extension programs associated with and led by the land-grant universities.

County Extension

County Cooperative Extension is the part of the national extension system that can be, and sometimes is, most attuned to the people and communities of the nation. While it is clear from the preceding chapter that it is not always sufficient to give an audience just what they want, it is also clear that programming close to the people being served has some of the best chance of making a difference. However, the history of successful technology transfer in extension and the dominance of the universities' roles in the introduction of new agricultural technology tend to belie that proposition. It is unlikely that hybrid corn or artificial insemination in dairy cattle would have been programs developed at the county level or for that matter even asked for by farmers. At the same time, the arguments made earlier about the contribution of engagement to science suggest that technology transfer is only one-half of the relationship necessary to achieve engagement and its positive effects on the university.

There are those who argue that county extension is really where the action is in the extension land-grant system, and others who argue that county extension is good—for some things and some audiences but that not all extension programming must or need go through county offices.

Control of the County Extension Agenda

County extension generally has the support of the people it serves and the people who have input to its programs. Sometimes the people county extension serves are a cross section of the community and sometimes they are not. The independent programming decisions made at the county level by county staff and extension councils probably have as much or more to do with the character of the overall extension portfolio as do decisions made at the state or federal levels of the partnership.

In many counties across the nation, the character of the extension program is locally determined and strongly supported by the university, if the university can figure out how to contribute. If the land-grant university can't figure out how to contribute, local staff will proceed to find the support wherever they can. Richard Nunnally is the director of the extension program in Chesterfield County, Virginia. He is a state employee but fully funded by his county and extension is a department in Chesterfield County government. Chesterfield County is just south of Richmond and has a large urban and suburban population. Nunnally's response to questions about how county staff decide what programs to conduct and just what their relationship is to the land-grant universities (Virginia Tech and Virginia State University) was:

> We DO plan the majority of our effort around LOCAL priorities. My understanding of our system has always been to utilize research-based information to meet LOCAL needs. We use Tech (Virginia Tech) and Virginia State as our first source of information, but go on to other universities as needed to meet the needs of our clients. I think my district director approves of all of our programs in this county. We do have a few that are unique because of the strong support we have from our local government.
>
> I work closely with the green industry and do training for garden center employees. We have a Learning Center for limited resource clients in a primarily Asian community. We are partnering with a foundation-funded project called YOUTH MATTERS to help develop strong families and build strong communities. We're coordinating an educational effort to promote quality childcare (funded by a grant). We, along with several other Urban units, are targeting residential water quality programs.
>
> All of the above address some priority ISSUE as recognized by the state and federal partner.... But most importantly, they meet defined LOCAL NEEDS (Nunnally 1999).

Richard Nunnally has described rather effectively the relationship between the land-grant university campus and the counties throughout the country. From the county perspective, if the university can't help, the county staff will find the help elsewhere. If the agenda of programs that the university and/or the federal government is promoting fits what the county staff see as genuine needs and issues in their county, then they are quite prepared to use those programs and/or report their work under those categories. But, county staff are very clear—they set the local agenda locally in consultation with local leaders.

A County/Campus Programming Disconnect

In Minnesota in 1999, people spoke of two extension services—one in the counties and one on the campus of the University of Minnesota. In Minnesota, as well as in many other states, there is a substantial disconnect between extension in the counties and the programs they deliver, and what is done by university professors with campus extension appointments.

This writer's own experience developing and implementing a local government financial management program for Massachusetts' cities and towns in the late 1970s and early 1980s was that we suffered the same disconnect. The principle employed in the local government program in Massachusetts was to develop the means to deliver educational programs directly to the local government audience in a way that did not depend on the county extension staff. This was done because there was either lack of interest or lack of confidence in dealing with a new subject by county staff, and we were unwilling to have the program wither before it even got to the audience. At the same time, an effort was made to make it possible for interested county staff to choose to involve themselves in the program and earn credit with their local communities for their efforts.

From the perspective of campus faculty, part of the programmatic disconnect comes about because field staff are not interested or do not feel comfortable in the subject matter that is being promoted by the campus faculty member. Yet others feel that county staff are too fractured in terms of their programming responsibilities to spend enough time in any particular subject matter to become sufficiently competent to contribute in the delivery of programs. In the case of agricultural issues, as the problems of farmers become ever more sophisticated and technically complex, the capacity of an agricultural generalist county agent to deal with some of the specialized problems declines. All of these campus perceptions of the source of the disconnect has led to ever more complex arrangements by extension in various states to reorganize itself to accommodate the need for greater specialization and competency while still maintaining the imagery of a county by county presence.

County Staff Specialization and Its Management. One such staffing arrangement is to organize a "cluster" of counties as a planning unit for extension programming. By having an array of various specializations within the cluster and by having staff working across county lines within the cluster of counties, the possibility exists of having extension staff housed in one county but working in their specialization across counties. In Virginia, such an arrangement appears to have improved the ties between campus faculty and field faculty. However, the same arrangement in Minnesota, by some accounts, has not worked as well.

In Minnesota, at least some field faculty have multiple specializations and some of those are not complements. For example, several field staff members have a split assignment between community development and 4-H. The 4-H work dominates and the campus-based economic development specialist says it is almost useless to try to get the time and attention of those individuals for any of his community development work (Morse 1999). Further, it appears that notwithstanding the enthusiastic support for the cluster arrangement at the level of Minnesota extension administration, the concept is not nearly as popular with the county extension advisory committees. Minnesota county staff reported that they did work outside of their county within the cluster but did not tell folks in their county about it because it would not elicit approval.[1]

Finally, in order for the full complement of skills in a cluster to be available to the people in any county, some further management and adjustments by all staff in the county are required. It was reported that in most Minnesota counties, the secretary answering the telephone in any county office only considers the staff in that county when fielding phone requests. By not considering all of the staff in the cluster to be a resource to the county, the fact and public perception will be that the only issues extension in the county can address are related to the skills resident in the county extension office. Telephone transfer systems, other inter-county office communications mechanisms, and clerical staff training within the cluster, besides the formal agreements to share skills, will be required to make such arrangements work.

That farm groups in many counties continue to oppose clustering or similar arrangements aimed at improving the level of agricultural skills available to them is one of the greater curiosities of the capturing of extension by agricultural audiences. It seems inexplicable that they would seek to retain an agricultural agent who is a generalist. There appear to be two plausible explanations. The first has to do with the one-on-one consultant-on-demand style that is an ever more common mode of operating for agricultural extension agents in many counties as farm numbers decline. Local farmers like that kind of attention, and may be fearful they will not get that kind of service under the changed arrangements. It is entirely possible that they would still get that attention under different management arrangements and from even more specialized agents.

A second plausible explanation has to do with the tendency of many in the agricultural extension program to be public advocates for their farmer clients. If you have a local spokesperson for agricultural interests who is one of the best educated people in the county and speaks with the authority of the university on your behalf, why would you want that to change, even if you know more about contemporary farming issues than he does?

Different Time Frames for County and Campus. From the county's perspective, one of the biggest problems associated with looking to the university for support is the time frame in which people live. University people, especially those without major obligations to support extension programs, live in a teaching semester time frame. Thus, when a request for assistance is made to a university faculty member for some support such as research or other development, the campus faculty member believes he is responsive if the reply is, "Yes, I can work on that next semester when I have a lighter teaching load." The county faculty/staff member considers that as unresponsive because they need it no more than five weeks from the time of request, if not by next week.

The time frame issue in universities is a general one that is relevant to all public service and even to some contract research/analysis for both the private and public sectors, not just to the support of extension programs. This "responsiveness" or time frame issue becomes particularly critical when breaking into new outreach/extension program areas where there is not a history of funded research and extension or of a research establishment available to handle the rapid turn-around to a pressing local problem. That is yet another explanation for the dominance of the existing extension program portfolio by the traditional production agricultural programs—it is in production agriculture that there is the greatest reservoir of funded research resources and knowledge available to respond rapidly to new questions.

In many of the other areas where extension programs are offered, there is plenty of university talent available to support programs but not nearly comparable levels of research resources available to support extension programming. Thus the time frame issue often defeats efforts to respond to the educational opportunity. Several of the programs described in Chapter 8 illustrate particularly promising directions for extension programming based on problem-solving research in the human sciences. They illustrate that when the scholarship is available it is possible to develop outstanding programs based on many different scholarly disciplines.

Timely response is slightly better when requests are made of campus extension specialists who have part-time or full-time extension appointments. The ideal arrangement occurs when there is a true collaboration between field staff and campus staff in program development. Such arrangements do engage the university resources and make both the outreach and the science/scholarship better. It sometimes happens but not enough to make a full county extension program agenda. And, county staff will tell you they cannot wait until the university gets its act together.

The loss to the land-grant extension system resulting from the necessity for county extension staff to initiate programs without university support and engagement is that many of the excellent programs thus developed are unique to particular counties and to the organizational skills of particular county educators. While many such programs may be highly applicable to other counties and areas of the country, they are often not documented or organized in ways that make them easily transferable. This suggests another source of disconnect between county staff and campus staff that deserves further explanation.

Campus Faculty Incentives Are Different. Just as any broadly based county extension program cannot depend on collaboration with just a single or even a few university faculty members for their program support, no extension faculty member at the university is waiting around to hear the phone ring from a single county office. For a university faculty member with an extension appointment, work in a single county does not make an extension program. For the faculty member with a major extension appointment, the issue becomes whether one should develop the knowledge, information, and educational program that will answer the question for county A and answer it for county B, C, and D as well. Further, if the answer is prepared in a way that serves A and is also available to serve B, C, and D, it may begin to look like something that will have currency in the academic realm as well. Mike Sikora, formerly of Plymouth County Extension in Massachusetts, was always telling this writer, "You're trying to build a Cadillac, George, and all I need is a Volkswagen." Mike represented County A, and this writer was thinking of counties B to G, and the department Promotion and Tenure Committee besides.

Overcoming the County/Campus Disconnect

There are two formal institutional arrangements with which this writer is familiar that explicitly deal with the disconnect between county staff and university staff. One is the arrangement being tried at Oregon State University and the other is the Area of Expertise Teams (AOEs) that have been organized at Michigan State University. The logic being employed by Oregon State University goes something like this:

Extension field staff in Oregon have faculty status. The new definition of scholarship and the companion position descriptions employed by the university to evaluate faculty for promotion and tenure make it possible to use the same system for field faculty as for campus faculty. With the common basis for evaluation in place, Oregon State University and Oregon State University Extension made a decision to integrate field faculty into campus departments and to hold campus academic units—departments and colleges—responsible for extension programming decisions as well as for field faculty performance evaluations. Field faculty members are assigned to a campus academic department of their choosing. Field faculty evaluations are carried out by that department and field faculty participate in the evaluation of their departmental peers. That closer relationship between field and campus faculty, it is argued, will lead to greater engagement between campus and field.

The Oregon State experiment will be very much worth watching in terms of its influence on both county programming and campus scholarship. Chapter 8 describes the process of change at Oregon State in much greater detail.

The Michigan State Area of Expertise Teams are described as follows on the Michigan State University Extension Web page:

> Michigan State University Extension has Area of Expertise Teams in the following subject areas: beef, Christmas trees, community development, consumer horticulture, dairy, economic development, equine, family resource management, farm management (FIRM), field crops, floriculture, food nutrition and health, food safety, forage/pasture/grazing, forestry, fruit, human development, land use, leadership (Lead Net), livestock, manure, ornamentals, poultry, sheep, state and local government, swine, tourism, turf-grass, vegetable, volunteer development, water quality, woody ornamentals, and youth development.

What are AOEs

(1) AOE teams have co-chairs; one from the campus and another from off-campus.
(2) AOE teams develop their own microvision and operating procedures.
(3) AOE teams have an interdisciplinary, problem-solving, customer-orientated focus.
(4) AOE teams develop a plan for program delivery and curricula for staff development.
(5) Involvement of stakeholders is expected, including stakeholder information input for program/project selection direction and evaluation.
(6) Each AOE Agent member has an opportunity to select a mentor.
(7) AOE teams are expected to be entrepreneurial and generate resources for enhanced programming.

Self-Directed Work Teams

Self-directed work teams are a group of employees who have day-to-day responsibility for managing themselves and the work they do with a minimum of direct supervision. Members of self-directed teams typically handle job assignments, plan and schedule work, make production and/or service related decisions, and take action on problems.

Membership Criteria

Within each AOE, a team will be configured consisting of provost-appointed staff and research and extension faculty. The team will be associated with appropriate academic department, institutes, centers, or programs on campus (MSUE 2000).

According to people in Michigan State University Extension, the AOE teams appear to work the best when the field faculty members are full-time in the subject matter of the teams and have expertise of their own to bring to the team. This appears to work the best in the agricultural specialties where there are field staff members with considerable expertise. Where the team members have minor assignment in the subject and/or no particular expertise in the subject matter, the teams suffer from the need to both develop programming and to bring team members up to speed in the subject matter as well. Campus faculty find it a burden to simultaneously be trying to train and program at the same time. Several of the AOE teams are filled with staff with less than 20 percent assignment in the subject matter area and they are proving problematical. At least one extension specialist expressed the view that no one should be permitted on an AOE team unless they had a 50 percent commitment to the area.

The AOE teams also carry an administrative load as well. Prior to the teams, there was a program leader for children, youth, and family, but now there are AOE teams in youth development and family resource management. By one perspective this is a way to force administrative costs on to campus extension faculty and to give the appearance of less administrative overhead than is reality.

Overcoming the campus/county disconnect is no easy task.

The County and Field Staff Context

The relationship between land-grant universities and county extension, and, in turn, the influences on county extension programming, are much more complex than simply the problems of individuals in different systems attempting to work together for mutual benefit. There is considerable variation in the relationships between county extension staff and county government; between county staff and county extension advisory groups; and between counties and the universities in the financial contribution from county governments to the extension program. These also have a bearing on county level programming.

The financial contribution of counties to extension budgets in 1997 varies from 36 percent in Florida and Colorado to zero in a number of states. Further, there are other differences in the involvement of counties in extension programming and guidance. In a number of states, some or all of extension staff in the counties are county employees. In other states, clerical staff in county extension offices are county employees, and in yet other states all of the county extension staff are university employees. Richard Nunnally, whose earlier reported insights to county programming were helpful, was quite emphatic that he is totally conscious that he is responsible for a department in county government, even though he is a university employee.

A farm management extension agent in Virginia spoke of some of the pressures from these various arrangements at the county level to an accrediting review team visiting the Department of Agricultural and Applied Economics in June 1999. The agent explained why it is that he spends considerably less than 100 percent of his time on farm management. The agent serves a group of counties—a planning district or sort of cluster arrangement. He gets about one-third of his salary from county sources and the rest from the state and federal dollars. Previously he was fully covered by state funds and was able to spend most of his time on farm management.

He explained that his assignment is to respond to the needs and directions of several counties, the state, and the federal government, and his pay comes from all of them. Each of the units of government wants its share. What he actually does, he said, is dominated by the need to respond to the pressures in the office where he is housed and from the other counties, to support local extension activities, whether or not it is farm management. The result, he explained, is that he does less farm management and has a lesser relationship with the Department of Agricultural Economics now than when he was totally covered by state monies. Previously, he could say he was obliged to concentrate on farm management (AAEc 1999).

The Local Political Context. Besides the employment relationship between county government and county extension staff, one of the other important influences on county extension programs is the character of the extension advisory committees, county extension committee, or county extension councils. In most states, most counties have such a committee but whether they help or hinder county programming depends on your point of view. From the perspective of this book, well-functioning county extension councils that are representative of all of the people in the county are critical to developing the political constituency for a broadly-based extension program portfolio.

In Wisconsin, the County Extension Committee (also called the County Agriculture and Extension Education Committee) is a subcommittee of the County Board of Supervisors and is thus comprised of publicly elected officials. In other places, the empowerment of the extension advisory committee is solely from the university and the Cooperative Extension Service. In New York, all county extension educators are employees of the County Extension Association, which is a private nonprofit organization in the county with a board of directors who serves as the county advisory group as well. In some places, there are restrictions on the character and representation of the membership and the terms of service on advisory committees, and in other places there are no such restrictions.

It was reported to this writer several years ago that the position of chairman of a county extension advisory committee in one Minnesota county was passed down (inherited) from the retired father to his son in a prominent farm family in the community. A long-time USDA employee reported that in one county, a school bus driver was always on the extension committee so that the 4-H program would have easier access to the school buses. Clearly, some of the dysfunction of county advisory committees comes about through the complicity if not collaboration of county extension staff in stacking the membership in the advisory committee. This is particularly true when there are no membership rules or term limitations and when extension staff find a willing, supportive, and influential client/patron. This type of hostage taking is not limited to the agricultural program or agricultural interests. Family development agents report being captured by the long entrenched "Homemakers Club" members who were more willing to have extension programs on microwave brownies for themselves than to encourage programs directed to the serious nutritional problems of the county's poor.

The New York state arrangement that has county extension operating as private nonprofit extension associations is not the norm and is unique to that state. However, in the very early days of extension in the counties, an association of farmers as the sponsor of extension was the norm. The New York arrangement represents an interesting bit of the history of county Cooperative Extension and its early efforts to provide civic education, social change, and a collective problem-solving structure to farm people that is worthy of a brief acknowledgement.

In the years following the Smith-Lever Act of 1914, monies were provided by the federal government to establish county level agents of the land-grant colleges of agriculture. According to Mancur Olson (1968) many state governments required that a county could receive government money for a county agent only if there was an organized expression of interest by farmers for the educational services an agent would provide. These "Farm Bureaus" became the county agents' initial audience and support group. Indeed, according to Olson (1968), in the early years of county extension, access to the information assistance of the county agent required membership in the Farm Bureaus—it was rumored that some agents sent membership dues bills and educational materials in the same envelope. "The farmer who joined had first call on the county agent's services: the farmer who did not, normally had last call, or no call at all" (Olsen 1968, 150). Consistent with the urging of collective action on their own behalf by agricultural educators, many Farm Bureaus established business operations for members, not limited to but including their very successful insurance business.

When the relationship between the business activities of County Farm Bureaus, the lobbying activities of their national federation, and the publicly funded educational efforts of federally appointed county agents being directed primarily to Farm Bureau members was challenged, the Farm Bureaus separated from the extension operation beginning with the "True-Howard" agreement in 1921 (Kile 1948). The County Extension Associations in New York state as private nonprofit organizations retain the early status of private nonprofit, nongovernmental organizations consistent with the character of the early Farm Bureaus in that state.

In Illinois, more than 20 percent of the 102 county extension offices are still co-located (located in the same facility) with the county Farm Bureau. This occurs, notwithstanding a memorandum of the Secretary of Agriculture that discourages, and possibly prohibits, such association with "organizations of farmers whose functions include the influencing of legislation affecting the activities of this department" (Benson 1954). There may be ambiguity about the authority of the USDA and Secretary of Agriculture Memorandum Number 1368 for employees of the University of Illinois Extension, though the memorandum explicitly states that county agents as joint federal-state employees are subject to its restrictions. However, where extension is located in the same facility as the Farm Bureau there is no question about who is in whose pocket, or for that matter, about whether extension is committed to serving a broad community of interests in the county. On asking Peter Bloome, Associate Director of Oregon State University Extension Service, about the culture of 4-H when he was growing up on a farm in Illinois, Bloome reported that he was never a member of 4-H. It was, he said, because his father was not a member of the Farm Bureau and thought that was a prerequisite to his children being in 4-H.

County and state Farm Bureaus and the National Farm Bureau Federation continue to be among the strongest supporters of agricultural extension. They also are part of the major impediments to a broader extension portfolio and a larger role for extension in the engagement of the land-grant universities with America into the 21st century.

The New York County Extension Associations and their staff illustrate yet another of the "cooperative" part of Cooperative Extension. All permanent staff in the New York county associations have federal appointments and the benefit packages that go with that. As we will see, there are still entanglements and relationships between the three levels of government responsible for the Cooperative Extension Service system—some appear to work better than others, and some appear to work not at all.

Hearing County Voices

In many states, the county advisory committees are organized into a state association or federation that acts on behalf of county extension with the state extension office and on behalf of the state extension efforts with state government and state legislatures. The influence of state associations varies considerably. As one can imagine, when the Wisconsin State Association of County Extension Committees speaks out on an issue, the state legislature pays considerable attention, since all members of that association, and of the local committees, are locally elected county supervisors, with their own political support base in the counties of the state. More importantly, all of the members of the Wisconsin County Extension Committees are members of the Wisconsin Association of Counties, and that group is reportedly even more effective on behalf of extension.

Some of the special characteristics of the Wisconsin extension program are likely attributable to the fact that the county extension committees are comprised of independently elected officials who must be reasonably representative of their total community rather than speaking for a particular interest group in the community. In most other states, the membership of extension advisory committees is appointed and members often represent some particular interest group in the community. In several states, there are no statewide associations, and in other states, they are of little political consequence on behalf of extension. Though Iowa's county extension committee members are elected in the normal electoral process and thus have some political base in the county, until recently they had no state level organization. The newly organized state level body appears to be finding its voice on behalf of extension in Iowa (Johnson 2000).

When leadership of land-grant universities—presidents, provosts, and chancellors—start to contemplate changes in the relationships between extension (by whatever name it is called) and the rest of the university, they discover, particularly if they have not been previously initiated into the system, that there are many administrative, fiscal, and political entanglements between extension on the campus and extension in the counties that confound and complicate their decision-making. Questions arise that cause them to ask, "Why do I have to consider the views of this guy in West Overshoe County in making an internal university decision?" Academic leaders already know their job is akin to herding cats, and then they are introduced to the county and extension lions and tigers. It is no wonder that some land-grant university leaders, like Paul Torgersen, while president at Virginia Tech, take the position that extension is simply not one of the things they want to deal with (Torgersen 1994).[2]

It is unlikely that solving the problem of the disconnect between county extension and the university will be accomplished without the involvement of top university leadership and the cooperation of county level leaders. It is also unlikely that extension—in the counties or at the university—will have a significant role in engaging the university with the people of the state unless the disconnect between the two is solved. This is quite aside from issues related to the breadth of the extension portfolio and how much of the university can be engaged. Finally, if university leadership considers the politics of extension too messy and too much trouble and will not engage county and state extension politics, it would seem to indicate that there is not much of a commitment by that leadership to encouraging the university to truly engage the society and the public.

USDA—The Federal Partner

The Role of the Federal Partner

The Smith-Lever Act of 1914 states the purpose of Cooperative Extension is "to aid in diffusing among the people of the United States useful and practical information on subjects relating to agriculture and home economics, and to encourage the application of the same . . . ." (Rasmussen 1989, Appendix D).

The Smith-Lever Act as amended through 1985, restates the purpose is "to aid in diffusing . . . subjects relating to agriculture, uses of solar energy with respect to agriculture, home economics, and rural energy . . ." (Rasmussen 1989, Appendix D).

In both these acts, the monies appropriated by the federal congress were to flow to land-grant colleges of the states through the Federal Extension Service acting on behalf of the Secretary of Agriculture, "subject to the furnishing of equivalent sums by the states" (Rassmussen 1989, Appendix D)—state matching funds. The monies currently appropriated under Smith-Lever are allocated to the states by the Cooperative State Research, Education, and Extension Service (CSREES) of the USDA via a formula. A portion of the almost $1 billion for research and extension appropriated by the U.S. Congress in the federal partnership with the land-grant research and extension mission, is retained by the USDA for the operation of the CSREES agency. It is thus reasonable to assume that the working relationship between USDA/CSREES and the land-grant universities would be based on a great shared interest in the success of the partnership.

Until 1994, there had always been a federal extension agency in the USDA separate from the Cooperative State Research Service (CSRS), which handled the land-grant federal research monies. There also was a small organization under the CSRS that administered a small amount of money in support of higher education instruction in the agricultural sciences and related subjects. These separate offices are now dissolved and all activities combined in the single agency, the Cooperative State Research, Education, and Extension Service. CSREES is organized around a number of functions, for example the administration of competitive grants, and some issues, for example natural resources. However, the CSREES organization eliminates any explicit extension function in its organizational form.

One of the most significant events in the diminution of the role of the Federal Extension Service occurred in 1979 when Secretary of Agriculture, Bob Bergland, combined the Extension Service (ES), the Agricultural Research Service (ARS), and the Cooperative State Research Service (CSRS) into the single Science and Education Administration (SEA). Prior to that time the Federal Extension Service as partner to the states always had its own administrator in direct line authority to the Secretary of Agriculture. With the establishment of the Science and Education Administration, the leadership of extension at the federal level was reduced to an assistant administrator and could only gain access to the Secretary of Agriculture through his or her administrator (Schaller 1999a). The further dissolution of an identifiable Federal Extension Service into the form now manifest in CSREES was much more easily accomplished given the changes made in 1979.

The following is what CSREES says about itself in 1999 on its Web page:

> The new Cooperative State Research, Education, and Extension Service (CSREES) is positioned for the 21st century as a dynamic change agent and international research and education network. CSREES expands the research and higher education functions of the former Cooperative State Research Service and the education and outreach functions of the former Extension Service. The result is better customer service and an enhanced ability to respond to national priorities.
>
> CSREES links the research and education programs of the U.S. Department of Agriculture and works with
>
> - Land-grant institutions in each state, territory, and the District of Columbia;
> - More than 130 colleges of agriculture, 59 agricultural experiment stations, 57 cooperative extension services;
> - 63 schools of forestry;
> - 16 1890 historically black land-grant institutions and Tuskegee University;
> - 27 colleges of veterinary medicine;
> - 42 schools and colleges of human sciences;

- 29 1994 Native American land-grant institutions;
- 190 Hispanic-Serving Institutions.

Mission

In cooperation with our partners and customers, CSREES provides the focus to advance a global system of research, extension and higher education in the food and agricultural sciences and related environmental and human sciences to benefit people, communities, and the Nation.

The CSREES mission emphasizes partnerships with the public and private sectors to maximize the effectiveness of limited resources. CSREES programs increase and provide access to scientific knowledge; strengthen the capabilities of land-grant and other institutions in research, extension, and higher education; increase access to and use of improved communication and network systems; and promote informed decision making by producers, families, communities, and other customers (CSREES 1999).

Debate continues as to whether the new combined agency serves extension interests as well as a separate extension agency. There are those who argue that the current staffing of CSREES is by people whose backgrounds are more oriented toward the research agenda and do not understand extension. Others argue that the dominant emphasis of the CSREES reflects the on-the-farm production technology and management bias discussed in Chapter 6—not a surprise, given that emphasis in the system generally. What does seem apparent is that there is and will continue to be a disparity between the extension agenda in the states and the agricultural research agenda in the states. That difference leads to conflicts between extension and agricultural research interests at the federal level.

The Federal Agenda—Who It Does and Doesn't Serve

The best profile of the federal agenda in research and extension in its partnership with the land-grant universities can be seen in the CSREES budget and the relative monies in the respective categories.

Table 7.1 was constructed by taking the published list of CSREES research budget items and then attempting to match the extension budget items to the research categories as closely as possible. In the FY 2000 CSREES budget, there is an additional $39.5 million in integrated activities where the split between research and extension is not identified. Table 7.2 shows the character and budget for the integrated activities. The total CSREES budget of $950 million is the sum of the research/higher education budget, the extension budget (Table 7.1), and the integrated activities (Table 7.2).

Table 7.1 Appropriated budget, CSREES/USDA, FY 2000

Research Programs	$ million	Extension Programs	$ million
Base Programs:		Base Programs:	
Hatch Act	180.5	Smith Lever Formula (3b& c)	276.5
McIntire-Stenmnis Coop Forestry	21.9	1890 Colleges & Tuskagee Ext	26.8
Evans-Allen Program	30.7		
Animal health and disease, s1433	5.1		
Special Research Grants:		Smith Lever 3(d) Programs:	
Pest mgmt. alternatives	1.6		
Expert IPM decision sup. system	0.2		
Critical issues	0.2	Farm safety	3.4
Global change, UV-B monitoring	1.0		
IPM	2.7	Pest management	10.8
Minor use animal drugs	0.6		
Nat'l bio. impact assess. pgm.	0.3		
Minor crop pest mgmt., IR-4	9.0		
Rural development centers	0.5	Rural development centers	0.9
All other special	57.7		
National Research Initiative:			
(Competitive Grants Program)			
Nat'l res. and environment	20.5		
Nut., food qual. and health	16.0		
Plants	41.0		
Animals	29.0		
Mkt., trade, and RD	4.6		
Processing and new products	8.2		
Other Research:			
Critical ag. materials	0.6		
Aquaculture centers	4.0		
Sustainable agriculture	8.0	Sustainable agriculture	3.3
Supp and alternative crops	0.7		
Higher education programs	27.1	Agriculture in the classroom	0.2
		Children, youth, and families at risk	9.0
		Expanded food and nut. ed.	58.7
		programs (low-income families)	
		Other Extension Programs	
		Renewable resources	3.2
		Rural health and safety	2.6
		1890 facilities	12.0
Research at 1994 institutions	0.5	Extension at 1994 institutions	3.4
		Ext Indian reservation program	1.7
Federal Administration	10.7	Federal administration	11.8
Total Research and Education	**486.5**	**Total Extension**	**438.0**

Source: U.S. Department of Agriculture, CSREES, http://www.reeusda.gov/budget/webfund.htm, March 1, 2000.

Of significant interest is the discrepancy between the research and the extension budgets, notwithstanding the combining of the functions within the one agency. Remnants of the earlier administrative setup are still evident in the budget presentation and perhaps reflect on the

Table 7.2 Integrated Research/Extension Activities, CSREES FY 2000

Programs	$ million
Water quality	13.0
Food safety	15.0
Pesticide impact assessment	4.5
Crops at risk from FQPA	1.0
FQPA risk mitigation for major crops	4.0
Methyl bromide transition program	2.0
Total, integrated activities	39.5

Source: U.S. Department of Agriculture, CSREES, http://www.reeusda.gov/budget/webfund.htm, March 1, 2000.

earlier discussion about whether the combined agency serves extension interests or not. The monies for agricultural higher education ($27 million of the $950 million) are presented as a subcategory of the research budget. This is essentially the way that the organization was structured prior to the establishment of CSREES—the higher education office was under the Cooperative State Research Service (CSRS).

Of even greater interest is the discrepancy between the extension issues and the research issues. Notwithstanding the rhetoric about integrating extension and research, which is supported by the $39 million in integrated programs of Table 7.2, the remaining agenda is in many ways quite different. Peruse the list of topics in Table 7.1 and consider the degree of agreement between the research and extension agenda at the federal level in 2000.

The 4-H program, the extension program that most commends extension to the people of the United States, is supported in the states with Smith-Lever formula funds. The $9 million for Children, Youth, and Families at Risk is a special initiative directed to particularly vulnerable kids in the society and is carried out in the states through the 4-H programs. There is no support in the research budget for the 4-H program or for the Children, Youth, and Families at Risk program. Notwithstanding all the kids in America who have been taught to raise calves and pets of all kinds, extension still knows more about the calves than the kids or the problems of kids in the society—a most fertile ground for research generally located in the colleges of human resources within the land-grant universities.

Similarly, there is no research support for the Expanded Food and Nutrition for Low Income Families (EFNEP), another of the few programs in the official federal extension portfolio that explicitly serves disenfranchised and unorganized people of our society. The research needed in support of EFNEP would center heavily on behavioral aspects of dietary and eating habits of various groups in the society as well as on human nutrition and food safety. This is another research domain explicitly in the human science colleges in the land-grant universities.

After viewing the FY 2000 budgets, it is clear that the only integration of the research and extension functions at the federal level are for programs directed to farming and farm/environment issues. The Children, Youth, and Families at Risk and the EFNEP programs of the federal extension agenda represent two of several activities that are moving extension to serve broader than agricultural audiences. It would be cynical to believe that the lack of a research budget in support of EFNEP and Youth and Families at Risk is because the main constituents of the USDA do not much care about the people to whom these programs are directed, and see the programs as $69 million that would otherwise accrue to serve farmers' needs.

It seems clear that there is little or no shared interest on the part of the USDA to in any way enhance the capacity of extension at the land-grant universities to broaden their outreach portfolios with the help of USDA funding. This is not a surprise—were it otherwise would indeed be a surprise. The point is that there is a discrepancy between the interests of the USDA in extension and the land-grant universities' interest in extension. Much of the discrepancy has to do with who each sees as its clients/customers/constituents. In the drive to broaden the extension portfolio, the interest in nonfarming audiences is increasingly important to the land-grant universities. The USDA, with good reason, is still substantially vested in the interests of its farming clientele. The ability in 1992 of the meat and cattle interest groups to cause the USDA to withdraw its newly released nutritional pyramid is a case in point. The nutritional pyramid was rereleased without any change almost a year later in 1993.

The States' Dealings with the Federal Partner

There are those who argue that from the perspective of the states, that what happens in CSREES does not matter much either way so long as the federal funds flow. Some accountability is required for the federal funds, but most states can write the federal reports from the information they need for their own internal management without much difficulty (Wadsworth 1999). In discussion with a senior CSREES staff member about the declining contribution of the federal partner to extension in the states, the author commented that the states did not salute the

USDA partner quite so smartly as they once did. The response was, "Oh yes, they still salute, but now it is with one finger."

The declining contribution of the federal partner is not the only reason the federal-state relationship has changed in character. It is reported that there was a period prior to the 1950s when representatives of state extension service directors were included in the process of preparing the USDA's budget request for extension programs (Foil 1999). In 1999, each of the state extension directors was still "approved" or appointed by the Secretary of Agriculture, and they were therefore "federal appointees." That courtesy involvement of federally appointed extension directors in the USDA budget process has long since gone and indeed, representatives of the land-grant universities now collectively prepare an alternative to the administration's request for both extension and research.

That alternative budget process is brokered through the extension (Extension Committee on Policy, ECOP) and experiment stations (Experiment Stations Committee on Policy, ESCOP) committees of NASULGC. Curiously, the draft of that alternative budget document for FY 2000 and several past years as well, has been prepared in the offices of the private lobbying firm, AESOP, Inc., rather than by the staff of NASULGC. The major influence on the alternative budget preparation is reputedly the deans of agriculture, with such input as any of the other interested land-grant entities are able to exert.

Because NASULGC, which is supposedly the major broker between the states and the states' several interests in their dealings with the federal partner, is a 501(c)(3) nonprofit entity, it is prohibited from formal lobbying. That is the reason for the employment of a private lobbyist. However, NASULGC also substantially defers the brokering between the extension and experiment station (research) interests to AESOP, Inc., as well. The final preparation of the alternative budget occurs in AESOP, Inc. NASULGC does play a role in brokering between the several states vis-à-vis extension interests within ECOP through the offices of Dr. Myron Johnsrud, the last Federal Director of Extension, now a NASULGC staff member.

The prominent role of the privately hired lobbyist in this process is a commentary on the character of the USDA/land-grant partnership, and on the effectiveness of NASULGC as broker to that relationship. While the monies for the lobbying contracts run through NASULGC, the relationships with AESOP, Inc., are with the several committees who have contracts rather than with NASULGC and its staff.

Dr. Rodney Foil, long-time dean of Forestry, director of the Agricultural Experiment Station, and vice president for Agriculture, Forestry, and Veterinary Medicine at Mississippi State University, suggests that among the reasons for the USDA's diminished interest in its role in the extension partnership is the result of frustrations of a number of Secretaries of Agriculture who discovered they did not have the ability to command the land-grant partners in the same way they could other of their staff (Foil 1999).

One such story involves Secretary of Agriculture Orville Freeman in the Kennedy administration. In 1963, Secretary Freeman and his economic advisors were seeking to have a referendum by farmers that would affirm acreage controls on wheat. In exchange for allowing small farmers in the fringe areas of wheat production to participate in the vote, a right they had not previously had, Freeman was expecting them to affirm the referendum. He expected "his" county agents, many of whom had federal appointments and the benefits that went with that, to rally to his cause and help sell the program to farmers. The state extension systems and the county agents were unwilling to advocate a particular policy outcome and instead carried out a rather more balanced public policy education program. The referendum was defeated. Freeman, who was reputed to be cool on extension anyway, was apparently put off by the experience (Tweeten 1999; Schaller 1999; Schertz 1999; and Schnittker 1999).

Others, most particularly John Schnittker (1999a), former Assistant Secretary of Agriculture, suggest that while the details of Freeman's miff on the wheat referendum are essentially true, the declining fortunes of the extension system, the declining federal contribution to extension, and the essentially "pass through" function of USDA to the states' extension budgets, explain more of the relationship between the state and federal partners, than do such incidents as the Freeman wheat referendum. In support of this view, several other long-term observers of the federal/state relationship suggest that the single finger salute was always present on the scene in the states' response to federal direction.

The Dilemma of Trying to Keep Everyone Happy

From the internal staff perspective of CSREES, where much of the funds they handle are pass-through monies and a smaller proportion are competitive grants, the major role of CSREES staff is to urge, cajole, facilitate communication, and share ideas with state research and extension staffs with respect to the federal agenda. There are those within the land-grant universities who argue that because the CSREES staff have little direct authority and have all of the constituents listed above, they have great difficulty taking hold of any idea or initiative and really making something happen on behalf of their land-grant partners.

There is a story on the street that in the fall of 1995, and again in January 1996, the National Corn Growers Association approached the CSREES staff with a proposal to seek support for a corn genome project to better understand the genetics of corn, one of the nation's most important crops. The corn growers were advocating a modification of the National Research Initiative (NRI) to increase the funding in it for corn genome work. The corn growers, it is said, were prepared to give up virtually all other research on corn in order to achieve the new emphasis they wanted in the NRI. They even suggested they would go after an additional $10 million for the NRI to achieve their goal of more corn genome research. CSREES leadership declined to consider any changed emphasis in the NRI, or to consider the proposed effort to raise additional funding for the NRI, on the grounds that they could not open the door to every other special interest group's special research agenda. In frustration, the corn growers approached the Agricultural Research Service, the USDA's own in-house agricultural science research agency, but were also turned down by that agency. Finally, the corn people went to Senator Bond of Missouri, who chairs the committee responsible for the National Science Foundation, for his advice and assistance.

It is true that in 1998 the National Science Foundation, which has been very aloof in its dealing with agricultural science and the agricultural science establishment, established a Plant Genome Project. In its two-year life, the project has put out grants to the tune of $110 million, of which $35.5 million went to research on maize (corn). Of the $110 million, about $68 million went to land-grant universities, and the balance to private or other public research institutes (NSF 1999).[3]

The point is that a research project whose time had come but was not able to be developed or encouraged by the CSREES staff, was so viable and politically supportable that it broke down barriers within the National Science Foundation. The way that the National Science Foundation dealt with the special interests issues, presumably a concern for them as well as for the USDA, was to generalize the original request into genome research on plants generally rather than just to corn. It's too bad that CSREES leadership did not have the insight to see that modification of the corn growers' proposal could produce an alternative that they could have then supported.

Another story speaks to the similar ambivalence of CSREES about its role vis-à-vis the land-grant universities. It is said that in about 1991 the Office of Management and Budget was seeking information about the numbers of scientific persons spending time on water quality and associated issues in order to fathom the magnitude and significance of scientific efforts that would be affected by federal funding on the subject. The request came to the USDA generally and was sent to several of its agencies. The Agricultural Research Service (ARS), the USDA's own technical research organization, reported that they had some specific number of research full-time equivalent (FTE) staff time in the hundreds committed to research on issues of water. The CSREES reported that they had only two persons who spent any time on water, forgetting that they represented and supported many hundreds of state land-grant university scientists and extension staff, many of whom were working on water issues, and who would indeed be affected by federal funding on the subject.

Disparate Interests among and between the Partners

The single-minded/single-purpose government agency issue is endemic in our American system of government and goes deeper than just the behavior of the leadership of the USDA. The first time the Youth At Risk budget item was proposed to the Agricultural Committees of the U.S. Congress in 1989, it was jointly supported at $10 million by the USDA, the OMB, and HUD. The ground work for the program had been done by the last Director of Extension, Myron Johnsrud, and his chief budget officer, Richard Rankin. The proposal was based on a trip to view the work of an extension agent in inner city Chicago. Rankin and Johnsrud took a collection of people from ES/USDA, USDA's Budget Office, and OMB staff who had been critical of 4-H on the grounds that it was not serving urban kids or kids at risk. When the proposal got to the Congress, Representative Jamie Whitten, then the chairman of the House Agricultural Committee and the strongest congressional advocate for urban 4-H, cut the amount virtually in half because he could not withstand the assault of the agricultural interests (Rankin 1999).

These same agricultural interest groups are primary clients of the deans of agriculture as well as of the USDA. The battles at land-grant universities over the control of extension involving deans of agriculture and those who wish to have extension be more centrally located in the university discussed in earlier chapters, are reflections of tensions between extension interests from the states (ECOP) and agricultural research interests in the states (ESCOP) with the USDA.

The simple fact that the federal partner to land-grant extension is the U.S. Department of Agriculture becomes an argument in questions about both the administrative control of extension in the universities and the character of extension programming. The fact of the USDA as federal partner of extension feeds the internecine struggles over control of extension in the universities. The symbolism of the USDA as partner encourages the national network of farm interest groups to support the notion of ownership of extension by agricultural interests at national, state, and county levels.

Close ties with agricultural clients should be a grand arrangement for extension since agricultural extension will likely remain a strong part of any extension program. It would be a grand arrangement if the sense of ownership by farming interests did not consider their claims and support to be exclusive of other partners and programming. Unfortunately, farm interests at federal, state, and county levels mostly see broadening of the extension portfolio as a threat to programs that serve them, rather than as a means to maintain programs that may not be otherwise sustainable in the long run.

Dr. Terry Nipp of AESOP, Inc., the primary lobbyist for the land-grant experiment state and extension interests in federal appropriations, reports that he has great difficulty getting lobbyists from the major farm groups like the American Farm Bureau Federation to join his efforts in support of legislation of importance to research and extension that does not directly impact farming activity. This occurs, notwithstanding the published policy positions of such as the Farm Bureau on rural schooling, rural health care, and other areas of life in rural America. It is not, reports Nipp (1999), that the agricultural interest groups are hostile to his activities on behalf of the land-grants in other parts of the federal budget—they simply are unwilling to spend their political capital there.

The land-grant universities' traditional constellation of colleges, or core land-grant colleges (the colleges of agriculture, the colleges of human sciences, the colleges of forestry and natural resources and the colleges of veterinary medicine), has research and extension portfolios that include monies from many federal sources besides just the USDA, much of it through competitive grants and contracts. The NASULGC Board on Human Sciences did an inventory of the 1998 funding portfolio of 40 of the most prominent colleges in the human sciences. Of $172 million in total funding, only $35 million was from USDA sources, and $49 million was from other federal agencies. Of the other federal agencies, monies from the Department of Health and Human Services equaled the amount from the USDA at $35 million (Board on Human Sciences 1999). Nipp's position is very clear. He sees his job as seeking to enhance the flow of all of those monies to the land-grant universities. That is not a function that CSREES plays any role in, and, as the corn genome research example makes clear, CSREES has some difficulty in imagining new initiatives on behalf of even agricultural science at the land-grant universities.

Alternative Federal Partnerships

At the present, the USDA is the only federal partner with which the land-grant universities collectively have a formal relationship. Other relationships with federal agencies must be negotiated state by state in multiple bilateral arrangements—a very uncoordinated and piecemeal set of arrangements that further distinguishes the strong or powerful states and state institutions, from the weaker ones. In these bilateral relationships, who is strong and weak frequently is a function of internal partisan politics in the U.S. Congress. As one contemplates the future of the land-grant universities and the role of extension in their outreach activities, it is not difficult to argue that the partnership with the USDA is a significant liability to broadening the extension portfolio and engaging more parts of the universities under extension. NASULGC seems unable to develop much in the way of meaningful alternatives for even internal system wide discussion. Consider then several alternatives to the current arrangement for Smith-Lever and other extension formula funds now administered by USDA/CSREES.

Alternative Scenario #1 for Federal Funding of Extension. Since major portions of the federal extension budget are more consistent with the missions of agencies besides the U.S. Department of Agriculture, this alternative proposes dividing the existing formula funds by the major program categories and making the federal agencies most attune to the program issues responsible for administering that portion of the monies. One can imagine, for example, dividing the funds by the 1992 commitments of resources reported in Table 1 of Chapter 5. Using the 1992 figures by major program areas, 47 percent of federal extension funds would be committed to agricultural programs and would remain in the USDA. Twenty-four percent of federal extension funds would be committed to family and consumer sciences and would be administered by the Department of Health and Human Services. The 22 percent for 4-H and Youth could be administered by the Department of Health and Human Services or the Department of Education, and the 7 percent in community and resource development could be administered by the Department of Commerce.

Under this scenario, the core land-grant colleges would have as many as four different federal partners—the USDA, the Department of Health and Human Services, the Department of Education, and the Department of Commerce. As these federal agencies came to understand the capacity and abilities of the land-grant universities and county extension, they might find an interest in augmenting the formula funds in order to improve programming accomplishments. Further, agricultural interests at both state and national levels, who have been solid supporters of extension, could once again work for the enrichment of agricultural extension programs and not be distracted by the internecine struggles over who will control extension. Deans of agriculture could be in charge of agricultural extension and the overall leadership of Cooperative Extension could be placed wherever university leadership determined was most propitious.

Alternative Scenario #2 for Federal Funding of Extension. A second alternative scenario deserves some discussion. Under this scenario, all of the federal extension monies would be administered by the National Science Foundation. This would be quite an interesting symbolism since the message would be conveyed to the NSF, to the land-grant universities, and to the science research community generally that the people of the nation have a claim on all of the research funded by the federal government.

While it is true that the NSF is not the only federal agency besides the USDA that funds research, the most notable other agency being the National Institutes of Health, the NSF is symbolically the major general research-funding agency. Further, in recent years, even the NSF has discovered the need to demonstrate the efficacy of its investments in research and has recently added an outreach requirement in its program of centers for the materials sciences. One of the requirements for funding of a materials science center at any university is:

> . . . a description of proposed activities in education, human resource development, and outreach; proposed collaborations with industry and/or other sectors; shared experimental facilities; international collaboration; and an outline of the proposed arrangements for administration and management of the Center (NSF 1999a, 7).

Such an arrangement would break the stranglehold that agricultural interests have on extension. It would legitimize extension/outreach/engagement in the universities by any and all recipients of NSF funding—NSF is a significant source of funding for U.S. universities including the land-grant universities—a matter of considerable importance as the universities seek to become engaged. As argued earlier, having the extension obligation provide an institutionalized test of workability and relevance might make NSF funded science better and more relevant, just as the extension influence has made agricultural science better and more relevant.

Alternative Scenario #3 for Federal Funding of Extension. A major part of the problem for the land-grant extension system in its pursuit of federal funding support beyond its agricultural portfolio is the transaction costs in dealing with federal programs that are not within the CSREES budget. Each state must initiate its own relationship and contractual arrangement. For example, for several years Ms. Linda Benning of the NASULGC staff has carried out an information cum brokering program for extension throughout the country, directing their attention to grant/contract monies for nutritional education associated with the food stamp program administered by the Food and Nutrition Service of the USDA. Individual states must make their own arrangements with that other part of the USDA that handles the food stamp program and is outside of the CSREES/land-grant partnership. By 1999, all but two of the states had entered into arrangements with the Food and Nutrition Service and have done so to the tune of much of the $74 million they administer for their nutrition education programs (Benning 2000).

Indeed, the same barrier is true for the other parts of the land-grant system in more easily gaining access to funds on behalf of the whole system where there is need or desire for nationwide coverage and where costs of doing business could be substantially reduced. There is a great need for a national institution that can act on behalf of the land-grant universities and accept appropriated and/or system-wide contract resources where nationwide program delivery is required.

Consider a legislatively created Corporation for Public University Outreach, which is an analog for the Corporation for Public Broadcasting. The CPUO would have as a part of its base funding perhaps $100-150 million from the existing Smith-Lever appropriations, leaving with the USDA and CSREES the part of the Smith-Lever funds that are substantially agricultural in their spending outcomes. The CPUO would be able to receive appropriated funds from other agencies such as the Department of Health and Human Services for programs like the Youth and Families at Risk; from the Veterans Administration for special programs to reach veterans, particularly in rural areas; from the Environmental Protection Agency for a multitude of water and other natural resource oriented educational programs; from the Department of Housing and Urban Development for the multitude of urban extension programs that now exist but are done without national coverage; and from the Department of Commerce for economic development initiatives, among others. In addition to the receiving and brokering of public funds, the Corporation for Public University Outreach could and would receive private sector and philanthropic funds on behalf of the system. Both private and public donors could impose a particular set of rules appropriate to their program interests for funding they pass through the corporation.

A major function of the corporation would be to assist funding agencies and organizations and the land-grant universities to systematically develop programs (as distinct from projects) that will accomplish a public purpose and be nationwide in scope. Thus one can imagine that, while all public universities might be eligible recipients of funds from the corporation, only a single proposal for any subject or program area would be accepted from any state, forcing the multiple public universities of the state to negotiate their respective roles prior to coming to the corporation for funding. Because of the significant distinction between programs and projects in terms of the capacity required to carry out each, the corporation would focus on multiple-year commitments for funding whereever possible, to assure some longevity in capacity. However, even "program" does not mean forever, and so the Corporation for Public University Outreach would assist in bringing some outreach efforts to a close, another problem the system now faces.

The public corporation would have a board of directors with broad and prominent representation from across the nation and across interest groups. Some have even suggested that a modification of the existing National 4-H Foundation might serve such a role. Such an entity should be of some interest to agricultural interests who would be relieved of the battles fought at the federal, state, and local levels over the spending pattern of the CSREES funds. With the addition of the CPUO as a federal partner of land-grant extension, agricultural interests could grow the funds in support of agricultural interests and seek to get the very best service the resources will permit, without distracting battles with other claimants of extension resources.

The Federal Partnership in Perspective

The picture here painted of the USDA/land-grant extension relationship at the turn of the century is one of rather disjointed and separate interests. Noting historical differences neither argues that things should be different now nor changes contemporary reality. It is, however, of some historical interest. For example, it is worth observing that the Department of Agriculture was established in 1862, the same year that the Morrill Act established the land-grant colleges that grew to become the land-grant universities of today. According to the USDA Web page, Lincoln dubbed the USDA "the people's department." "In Lincoln's day, 90 percent of the 'people' were farmers who needed good seed and good information to grow their crops" (USDA 1999a). While the number the USDA uses on its Web page is wrong—really only about 50 percent of the population was farming in 1862 according to Drabenstott (1999)—today it is less than 2 percent. That fact, in and of itself, explains much of the change in the organizational relationships over time.

The land-grant universities have a social contract to be people's universities and the United States Department of Agriculture has an obligation to assure a safe and secure food supply for the country. As the demography of the nation changed and productivity in agriculture increased, it can be argued that the USDA strayed less from its societal obligations than have the land-grant universities from theirs. This is true, notwithstanding the USDA's complicity in the hostage taking of the extension function at the land-grant universities by agricultural interests. The USDA clearly bears some responsibility in suborning extension's role in leading the engagement of the whole university in the society as the land-grant universities seek to fulfill their social contract with the people of America.

Indeed, in earlier years the discussion of the portfolio of extension in the states was of some interest to leadership at the USDA. In 1961, Paul Miller, then provost of Michigan State University, addressed the Centennial Convocation of the American Association of Land-Grant Colleges and State Universities (the organization now known as NASULGC). Miller (1961) quoted a statement made by an Assistant Secretary of Agriculture at the same association's 1911 meetings as follows: "This association should not forget the great importance of other than agricultural lines of endeavor. There are twice as many people in vocations other than agriculture as there are in agriculture; and about half our people are directly interested in home economics. Why narrow this question to one of agriculture?"

But "narrowing the question to one of agriculture" has been precisely the character of the relationship between the land-grant universities and the USDA. It is what imperils Cooperative Extension's opportunity to play a role in engaging the land-grant universities in the society and it is what imperils Cooperative Extension's own future in the 21st century.

Conclusions

There is an African saying from Kenya that observes, "When elephants fight, it is the grass that gets trampled." Unfortunately, much of the evidence on the relationships between the partners of extension looks a lot like trampled grass. Some of the trampled grass is the spirit of staff in all levels of the partnership. Some is the confusion and disarray of administrative structures at land-grant universities about what is "outreach," what is "extension," and "why do we need outreach if we have extension?" Some of the trampled grass are the hard-working and underappreciated federal CSREES employees who strive to be responsive to their state and county colleagues, strive to take a larger view about the system, and are frequently ignored or marginalized without any kind of salute.

Some of the trampled grass are the extension directors and other university leaders who have lost their positions, in part because they attempted to move the extension organization toward a broader program portfolio and away from domination by the agricultural part of the program. There have been carcasses of such people in the 1990s at the University of Minnesota, Michigan State University, Clemson University, Virginia Polytechnic Institute and State University, the University of Missouri, the University of Alabama, the University of Georgia, Iowa State University, the University of Illinois, and West Virginia University among others. Curiously, a significant number of the carcasses were women—proportionately more than their representation in leadership roles.

The point is that the politics of the control of extension and the character of its portfolio at the county, state, and national levels is the stuff of rough politics and continues to be unresolved. The forces at work are the following:

- the immutable fact of the changing character of the society;
- a staff of professionals throughout the extension system, most of whom are dedicated to excellence in programming that is responsive to the people in that changed society;
- the traditional audience of extension who are resistant to the control of extension moving out of the colleges of agriculture and to changes in the program portfolio; and
- the increasingly fragile funding support for extension at all levels of the partnership.

The politics get played out at all levels of the partnership of the Cooperative Extension Service. It is played out in county extension advisory councils, in county government, in county farm groups, and in county extension offices. It is played out on land-grant campuses in the central administrations and in the colleges, in state farm organizations, and in state legislatures. It is played out in the USDA and on Capitol Hill in Washington, D.C.

From one perspective, the politics of Cooperative Extension are the stuff of American democracy. The people are speaking and the outcome is as it should be. That position is akin to the argument made by some economists and their followers who say that in a free market economy, the only people who are poor, are poor because they are lazy or stupid. They are earning what they are able to contribute and they have little to contribute. The rich are rich because they earn it by their contributions to the society.

In the case of the politics of Cooperative Extension, the continued domination by agricultural interests in both control issues and in program issues has more to do with failures in democracy than with the success of democratic processes. If by "Cooperative Extension" is meant that the people will speak and provide support for the programs they want, there is an abundance of evidence that suggests that Cooperative Extension still mostly serves the very same people that it served when established in 1914. That is not "the people" speaking. This evidence argues that in politics, institutional structures best serve those who established them, and institutional history may be more important for its residues of power than for its ideas. By that argument, Cooperative Extension is not very cooperative.

Notes

1. Coffee time conversation at regional extension meeting, Austin, Minnesota , April, 14, 1999.

2. In fairness it must be acknowledged that President Torgersen strongly defended Virginia Cooperative Extension's state appropriated budget against assaults from state legislators several years later into his administration and after the cited conversation.

3. These numbers were determined by simply summing the approved projects published by the NSF Plant Genome project as of July 15, 1999. Most of the awards were for multiple years and totaled the $110 million reported above. Because of the multiple year grants the $110 million amount is not an annual flow but a total of approved multiyear grants. It was subsequently reported by a source close to the Corn Growers that there was a 1998 allocation of $40 million of new money, a 1999 allocation of $50 million new money, along with about $20 million annually of established NSF plant science monies that would have been otherwise used in some connection to plant genome work, leading one to conclude that the annual commitment is in the order of $60 million.

8

Promises and Possibilities

Introduction

Thus far the analysis and arguments about extension and the land-grant universities have focused on the dysfunction that mitigates against extension playing a significant role in the outreach of the university or that mitigates the university becoming engaged with the people of America into the 21st century. Dr. Paul Miller[1], on reading draft material of this book's first six chapters, said, "So, now that you have me almost giving up on this struggle, which began for me when I was 11 in 4-H, in 1928, how do you glean out of what you have already said, and add further steps of positive vision and possibility to keep the extension and land-grant ideas vibrant and expanding?" (Miller, 1999).

This chapter provides evidence that some positive things are happening within land-grant universities and extension that are counter to the forces that threaten the system as described in earlier chapters. Some readers will react that the evidence given here is too little too late, given the preceding critique. Others will know of other significant and promising changes in particular places in the land-grant/extension system and argue that there is much more out there that could be described. To all of the fine extension people who are waging their own personal revolutions, in spite of the incentives within the system, and whose good works should be chronicled here but are not, the author offers his apologies.

Several of the programs or arrangements described are carried out by land-grant universities but are not a part of Cooperative Extension. They are chronicled here to demonstrate that there is a tradition within the land-grant universities that is beyond the control of contemporary extension. In some states, extension administration is not widely embracing and encouraging of outreach and engagement across the whole university. The large amount of outreach activity from land-grant universities not associated with Cooperative Extension makes clear that in many places outreach will happen with or without the encouragement of extension. In some places, those nonextension outreach activities will leave Cooperative Extension in the dust.

The major point of this chapter is to identify promises for the future of extension and land-grant universities. Therefore, considerable attention is devoted to institutional arrangements that seem to have overcome some of the barriers that mitigate against universities becoming engaged, or that mitigate against extension playing a leadership role in that engagement.

Some of the program examples described are chosen simply because they speak of excellence in programming or employ a different model than the technology transfer/expert answers model. Several are in the humanities. Most are scholarship driven but not in the physical, biological, or agricultural sciences. The history of excellence in programming in the agricultural sciences does not need repeating here. However, excluding discussions of agricultural programs here does not in any way imply that excellence in agricultural programming is not important or does not exist. Indeed, the extension land-grant system already knows how to do it well.

Moving Minds—Humanities Extension/Publications Program

In the late 1970s, Dean Bob Tilman of North Carolina State's College of Humanities and Social Sciences saw that founding an extension program in the humanities and social sciences would enhance the role of the college both on campus and across the state.

> He (Tilman) worked with the National Endowment for the Humanities (NEH) to fund a unique statewide program in which this college and the university's agricultural extension service collaborated in bringing four-session, public, free seminars to the citizens of North Carolina. Selected professors from the College faculty were employed to lead the first and fourth sessions on topics such as "First Amendment Freedoms," "Charles Dickens," or "The Small town in American Literature." Extension agents at the county level booked meeting rooms, supplied refreshments, and registered local participants. Using outlines and original videotapes supplied by the humanities extension, local discussion leaders selected by the county agents led sessions two and three of those early seminars, the primary object of which was to "move" individual minds. The program was the first of its kind in the nation (Clark 1999, 5).

James Clark, director of the Humanities/Publications Program at NC State, writes that the participants in the early seminars were adults of all ages but particularly prominent were public school teachers. The teachers found the content of the seminars and the supporting reading and video materials superior to the usual in-service training they were exposed to by local school systems for recertification. Teachers called for the Humanities Extension Program to provide certificate renewal credit to teachers who completed the seminars. Next came the requests for access to the materials themselves for use in the classrooms, then for seminar faculty to speak to classes of public school students, and finally for special in-service workshops for teachers as well. "Simultaneously, funding for this popular new extension program had been added to the University's state appropriations, and the NEH designated it as a national model. Success created mountains of work" (Clark 1999, 6).

James Clark tells of the further development of Humanities Extension:

> While increasing the number of topics offered to general audiences through public seminars, Humanities Extension also responded to the urgent requests from teachers. It set up an at-cost curriculum materials service for its print and video productions; established a newsletter called "We're Your Place"; and founded, with corporate support from Glaxo and Burroughs-Wellcome, a public school OUTREACH program. By the mid-1980s, these developments were dramatically increasing the College's presence across North Carolina among the public schools and the general public. The curriculum of public seminars now included more than forty topics, and the annual census of separate seminars and teacher workshops reached adults in all of the state's one hundred counties. OUTREACH classroom visits by college faculty have averaged over three hundred each school year and in 1998–99 rose to 570.
>
> Our annual summer writing camps for middle school boys and girls were first organized by me in 1987 in cooperation with county 4-H agents. These activities spread statewide in the 1990s and exerted a direct influence on the state's (4-H) camping curriculum as writing became a regular activity for campers. Since 1997, Humanities Extension/Publications has earmarked royalty income to make grants to county and state-level camps to support writing instruction and also to fund the development of a standard curriculum for this purpose (Clark 1999, 6–7).

It has been during the past decade that the Humanities Extension Program has evolved into the Humanities Extension/Publications Program and perhaps accomplished its greatest achievements as a program of university outreach. Dr. Clark continues the story:

> With the falling apart of the Soviet Union, Humanities Extension Co-Director Joseph P. Mastro, a Sovietologist, saw that both his academic discipline and the social studies textbooks about Europe and Asia for public schools students would be out of touch with reality.

Working with me and other campus colleagues and a group of consultant teachers from across North Carolina, Dr. Mastro created an up-to-date sixth-grade social studies textbook entitled, *Living in Europe and Eurasia*. It was adopted by the North Carolina Textbook Commission in 1992 and became the preferred social studies text in virtually every sixth grade classroom in North Carolina. Thirteen original videos shot on location abroad . . . supplemented the text.

In March 1993, the North Carolina General Assembly acknowledged the success of this ambitious engagement . . . by mandating that Humanities Extension now produce new social studies textbooks and videos for grades 4, 5, and 7 as well as a new edition of the sixth-grade text for the next state adoption cycle in 1997. Seed money accompanied this legislative mandate, and work on the new books and tapes began immediately . . . .

Professor Mastro died unexpectedly in December 1993 of a heart attack. He was 52. His ambitious project immediately became an inspiring memorial embraced by me as his co-director and by Dr. Burton F. Beers, the professor of history and vintage textbook veteran who became chief executive editor of the new four-book series to be called "Living in Our World."

The word Publications was added to Humanities extension at this time as new consultant teachers, faculty, and staff worked with us and University administration to meet the 1997 deadlines for the completion of the four mandated books. Editors Chris Garcia and Gail Chesson joined the effort as news about the exciting developments spread into the classrooms of the state and into the world of corporate publishing. As a result, unprecedented teacher in-service workshops for the wise use of the anticipated books and tapes began while the writing and editing were still under way.

Acknowledging the superior design and substance of our texts, the School Division of Macmillian/McGraw-Hill sought and got a license for the sale and distribution of the "Living in Our World" series in the state rather than compete with North Carolina State University, the copyright holder, for the in-state market at grade-levels 4,5,6, and 7.

Our partnership with corporate publishing, with selected public school consultant teachers, and within the University enabled the College of Humanities and Social Sciences, led by Dean Margaret Zahn . . . to meet the legislative mandate, win adoption, and sell over $11,000,000 worth of the new textbooks in North Carolina. Schools in every system in the state bought classroom sets of at least one of our new books; two-thirds of the systems bought sets of all four texts! Through them over 320,000 boys and girls are studying our world. And our production schedule set about making new supplemental videotapes available at all grade levels.

In public educational terms, what is the meaning of the realization of Professor Mastro's extension/publications vision as it materialized under my direction and the editorial wizardry of Professor Beers and our staff?

Some of the answers we already know. For the first time in the history of North Carolina, adopted social studies textbooks for grades 4, 5, 6, and 7 meet exactly the standard course of study set forth by the State Board of Education. In other words, global economics and geography in the context of cultural diversity are presented at grade level for our children who live in our world. What Humanities Extension/Publications has produced and put into classroom service in North Carolina in this decade has not been done by another university in any other state. And we have cash reserves to produce new editions as well as new books for other grades or subjects (Clark 1999, 2–4).

The North Carolina State University College of Humanities and Social Sciences is very proud of its extension program. It seems that the Humanities Extension/Publications Program at NC State is a natural partner to the Cooperative Extension program of the state, but they are separate, virtually unrelated entities, except for the contributions of humanities extension to field level Cooperative Extension programs. Says Jim Clark of the relationship between the two, "We are not at war and do not even have arguments. We simply do not live in the same house" (Clark 1999a). The humanities extension methodology, including delivering programs at the local level, is identical to Cooperative Extension's. Indeed, early humanities extension programs were initially offered in collaboration with county Cooperative Extension offices with Cooperative Extension staff as presenters. Humanities extension's writing program for youth continues to make unsolicited contributions to the North Carolina 4-H program and has committed book royalties to that purpose.

The story of the publications development of the Humanities Extension Program sounds remarkably like many successful Cooperative Extension programs around the country. Pesticide application handbooks and pesticide applicator training, dairy herd improvement associations, soil testing, and a number of the farm record and analysis activities have been developed by Cooperative Extension and have similar dimensions. Such programs are sometimes spun off into separate business entities and sometimes kept within the program depending on the parentage of the program, and other administrative vagaries. The Parents Forever program in Minnesota, described later in this chapter, has some of the same "life of its own."

The closeness of the humanities extension modus operandi to that of Cooperative Extension, and indeed its benevolence toward Cooperative Extension, comes from the personal emotional ties to extension of Dr. James Clark. He writes:

In looking back over my more than three decades of faculty engagement at North Carolina State University, I cherish the actions little and large that have allowed me to spend my career on the campus that I first knew as a young 4-H member from a very rural county. Success in club projects enabled me to attend the University of North Carolina at Chapel Hill on a national 4-H scholarship. During and after completing graduate work in English at Duke University, I rose in the ranks here at State where the assumptions of public service were both expected and rewarded. I never applied for a job elsewhere. Here I have won recognition for my teaching and research as well as for my extension work. I am as fulfilled working among citizens in the counties of the state as I am teaching students in the seminar rooms of the campus (Clark 1999, 7–8).

And yet, despite this strong affinity between Cooperative Extension and the Humanities Extension/Publications program at NC State, they are not together. They both seek to bring credit to themselves and to NC State University but they do not do it in concert with each other, nor do they ride each other's coat tails. Each has a separate political constituency in the state and separate campus allegiances. Dr. Clark suggests that the separation and bureaucratic antipathy between the programs is because each sponsoring entity on the campus, the College of Agriculture and Life Sciences, and the College of Humanities and Social Sciences respectively, is jealous of its own prerogatives, finances, and identity before the people of the state.

Nothing of which this writer is aware about shared administration, coincidental planning and programming, or joint political lobbying/education requires the loss of identity or funding as a prerequisite condition. Often the opposite is the case where university leadership seeks to impress the people of the state and their political representatives by demonstrating how many different entities from the people's university work on behalf of the people throughout the state. Indeed, the staff of Cooperative Extension in NC State, as is routine throughout the country, have prepared "Cooperative Extension Success Stories," which are tailored to each county and political district in the state. The technique is smart and appropriate. But there is so much more to tell if they brag on all the things North Carolina State University Extension has going on in the counties or congressional districts or state senatorial districts.

At North Carolina State University, it seems particularly tragic that there is not greater collaboration cum coordination on shared destinies. The campus seems alive with the kind of commitment to the university being engaged as expressed above by Jim Clark. Indeed, the following are just the more prominent separate entities as formal outreach programs from the campus:

- Cooperative Extension (second largest in the nation)
- Humanities Extension/Publications
- Industrial Extension Service
- Textile Extension
- The Science House (College of Physical and Mathematical Sciences)
- Center for Universal Design (School of Design effort to provide design support to physically challenged people)

Except for the monthly Extension Operations Council meeting that brings together senior staff from all of the academic units on campus as well as from the University Library and the Continuing Studies unit, there is little coordination or collaboration between the entities.

Everette Prosise (1999), assistant vice-chancellor and coordinator of the University Extension Network, reports that the members of the council talk about the value of collaboration. However, in the end, Cooperative Extension, which is the largest organization in the council, is afraid it will have its money used by someone else, or the others are afraid they will lose their own identity because of the Cooperative Extension insistence that such initiatives be CES-led and credited.

The bureaucratic paranoia of North Carolina Cooperative Extension and the College of Agricultural Sciences about the control over and character of its programs is in strong evidence. It was reported (Prosise 1999) that in recent years, agricultural interests in the state had expressed concern that initiatives by some of the home economists in obtaining grants and contracts to carry out educational programs to serve new or special audiences were diluting the dominance of the agricultural portfolio of "their" Cooperative Extension Service. Jim Clark knows when he leaves the Humanities Extension/Publications Program leadership that there will be fewer resources put into the 4-H writing program. Clark supports the writing program for reasons of the heart and because it's a good program, despite little acknowledgement by campus leadership of Cooperative Extension or the College of Agricultural and Life Sciences of the importance of it in their program.

What the Humanities Extension/Publications Program at North Carolina State University does demonstrate is that there is among the people of the nation a hunger for learning and knowledge in many aspects of their lives. There are many ways that universities can engage our people that are more than technology transfer, however important that is in our technological world. That "Charles Dickens," "First Amendment Freedoms," and "The Small Town in American Literature" were successful humanities extension programs in rural North Carolina is evidence enough that extension/outreach/engagement is about "moving minds" in many different directions.

Redefining Scholarship and Integrating Extension Field Faculty

The institutional changes at Oregon State University have been mentioned or alluded to several times as being particularly worthy to the end of making that land-grant university more effectively engaged in the lives of the people of Oregon. The changes under way there are particularly promising in light of the discussions of this book. The fundamental institutional arrangements of interest are, in brief:

- The new definition of scholarship applied to all members of faculty within the university as follows, "Scholarship is original intellectual work which is communicated and the significance is validated by peers. Scholarship may emerge from teaching, research, or other responsibilities. Scholarship may take many forms including, but not limited to: research contributing to a body of knowledge; development of new technology, materials, or methods; integration of knowledge or technology leading to new interpretations or applications; creation and interpretation in the arts" (OSU 1999).
- The requirement of faculty that all promotion and tenure decisions in the university will be based on an individualized position description that makes clear their assigned duties and expectations of them and sets forth the character of the scholarship against which they will be adjudged to be worthy of promotion and tenure.
- The integrating of all field faculty into academic departments on the campus.

Two other arrangements are integral to the successful functioning of the changes identified above, namely:

- the existing status of field staff in county extension offices as university faculty and
- having the leader of extension as a university-wide officer.

Both are particularly important to the effective integration of field faculty into academic departments. While the importance of faculty status for field staff is almost obvious, of equal importance but perhaps less obvious, is the status of the Director of Extension. The status of that position as having a university-wide responsibility makes clear the lines of authority and relationship between extension and academic units. To that end, the position of Director of Extension at Oregon State University, which had been subordinate to the Dean of Agricultural Sciences, was made a dean with equal status to other deans of the university during the academic year 1993-94 and was effective January 1, 1995.

It should be noted that two deans of Agricultural Sciences supported the change and worked hard to make it work. Dean Conrad J. (Bud) Weiser and his administrative team recommended that extension be administered at the university level. They convened a forum with agricultural clientele leaders and President John Byrne to advocate this change. Dean Thayne Dutson played a key role both in advocating the change in his former role as associate dean and director of the Agricultural Experiment Station, and in implementing the change in the College when he succeeded Weiser as dean. This is a unique posture for deans of agriculture at land-grant universities since they have almost uniformly opposed such moves. The recommendation for the change in extension administrative locus was formally made in a 1993 report, "On the University's Third Mission: Extended Education," prepared for then President John Byrne by a distinguished emeritus professor at Oregon State University, Emery N. Castle (1993).

Castle's report on Extended Education (1993) also recommended that extension programs be administered through academic colleges rather than directly from the central extension administration, and that academic departments become an integral part of extension programs. Though Castle did not explicitly recommend that field faculty in extension should be integrated into academic departments, he suggested such arrangements should be possible if individuals and certain departments chose to do so. His strong recommendation about the involvement of academic departments in extension programs paved the way for formal integration. President John Byrne made the decision to integrate faculty into academic departments in 1993. When he announced the decision he declared, "All extension service faculty, county agents as well as specialists, will be assigned academic colleges, and will have an academic appointment in the appropriate college . . . considering the faculty member's work assignment, academic training, experience, and, most importantly, individual choice (mutual agreement between the individual and the college)" (Olsen and Boyer 1999).

At about the same time that Castle was working on his extended education report, Dr. Weiser, who had been department head in horticulture, was serving as dean of the College of Agricultural Sciences. During his service as dean from 1991 to 1993, Weiser initiated faculty discussions that led his college to adopt a broader view of scholarship and job descriptions for evaluating faculty. He reports that while greatly influenced by Ernest Boyer's *Scholarship Reconsidered* (1990), the Boyer work and efforts of the Carnegie Foundation were focused on the rewarding of teaching. Dr. Weiser was additionally dealing with the problems of evaluating and rewarding extension scholarship and faculty contributions to team efforts (Weiser 1999).

The College of Agricultural Sciences put into practice a new college approach to promotion and tenure in 1994 under Dean Dutson. During this same time, a campuswide Faculty Senate committee devoted a year to intensive study and revisions of the University Promotion and Tenure Guidelines. The proposed guidelines, including the new definition of scholarship and the required use of the position description, were approved by the Faculty Senate unanimously in the spring of 1995. They were adopted by University President John Byrne in June of 1995.

To assess the impact of the new guidelines on the OSU campus, Dr. Leslie D. Burns, a member of the OSU Faculty Senate committee that developed the new promotion and tenure guidelines, carried out intensive interviews with campus faculty members and administrators in the fall of 1998 and winter of 1999 (Burns 1999). According to Dr. Burns, there were nine emergent themes resulting from her interviews:

- the definition of "scholarship" within the 1995 P&T guidelines allows for the acknowledgement of diversity of intellect and skills among faculty at OSU;
- standards of excellence within units have stayed the same or have increased;
- position descriptions can provide clear expectations for faculty;
- faculty now may have greater freedom to pursue scholarly endeavors related to teaching, technology transfer, outreach, and applied problem solving;
- impact of the guidelines has not been uniform across the campus;
- standards and expectations within disciplines sometimes may override university standards or expectations;
- accreditation standards sometimes may override university standards or expectations;
- lack of standardization of documentation of scholarship outside the traditional peer-reviewed journal article is problematic; and
- the need for on-going campus conversations regarding scholarship and the promotion and tenure process was identified (Burns 1999).

These emerging themes from the changed promotion and tenure procedures at Oregon State University are for the most part promising. The scholarly associations' influence through control of journals and accreditation standards is not at all surprising. There is a convention in disciplinary departments of accepting the publication of an article in a particular journal as evidence of scholarship, frequently without ever reading it (Lewis 1975). This means that any scholarship resulting in journal articles will still have advantage in the evaluation process over other kinds of scholarly expression where there is not a similar convention of evaluation. The limited control that universities have over the standards of scholarship and the coin of the scholarly realm has already been discussed in Chapter 3. That being the case, the new definition of scholarship and the requirement for position descriptions before the university will process promotion or tenure proposals seems to have started some change in the culture of scholarship at Oregon State University. There is promise that the change will reward and legitimize a broader array of scholarly endeavor than was previously the case.

The integration of extension field faculty into academic departments was really a separate action from the change in the procedures for promotion and tenure as the chronology of the various events makes clear. But it is also clear that the integration would have been much more difficult had the new guidelines not been adopted and employed, and that leadership of Oregon State had the sequence in mind. In a July 1999 presentation by Jeff Olsen, a horticultural agent in McMinnville, Oregon, and Charles Boyer, the head of the horticulture department, go to some length to describe the change in the definition of scholarship and the role of position descriptions in the university as central to the integration process. The Department of Horticulture at OSU was one of the departments most affected when 24 field faculty members were added to an existing roster of 34 for a total of 58. Implementation of the decision to integrate was completed in 1995.

According to Olsen and Boyer (1999) among the major issues that must be addressed in the process of integration are:

- assuring full faculty participation in decision making;
- building departmental community;
- increasing communication;
- refining the evaluation of scholarship; and
- geographic distribution of field faculty throughout the state makes campus meetings challenging to regularly attend.

When speaking about the way in which field faculty fared in the integrated department, Olsen and Boyer give the following results since 1995:

- one assistant professor (with MS) was promoted to associated professor with tenure;
- one associate professor (with Ph.D.) was granted tenure; and
- three associate professors (with MS) were promoted to full professors.

Olsen and Boyer (1999) conclude that extension field faculty members have been successful in the promotion and tenure process in the integrated department. The dean of extension agrees and points out that in addition, the integrated P & T process has appropriately denied promotion to two field faculty members in the horticulture department, also an important outcome that illustrates the continuing rigor of the integrated process (Houglum 1999).

Some early indications of positive influences resulting from the integration in the horticulture department are that of 11 new hires in that department since 1994, six of the 11 have been for extension field positions and three of the six have been candidates with Ph.D. degrees. The department head reports that the formal involvement with the academic department was critical in attracting more highly qualified individuals. It remains to be seen whether those with Ph.D. degrees will be better field extension educators.

In Chapter 4, the following reasons or incentives were given for why extension faculty might not much use the written word in their work and fail to work in a proactive mode:

- Time saving—if you can get away with winging it—why not?
- Self-preservation—when information is particularized to a user via a personal consultative type of relationship, the first and primary source to which the information is attributed by the user is to the person of the extension specialist, not the institution he represents. Extension specialists use that proclivity by clients associated with personalized distribution of extension information to build direct personal political support.
- Avoidance of scrutiny—if you don't write it down, it is a lot easier to get away with fuzzy economics, biology, or engineering, actual misinformation, and/or undefended opinion.
- Frustration—if you can't get scholarly credit for it anyway, why bother?

One can reasonably speculate that the redefinition of scholarship and the integration of field faculty into departments with its concomitant pressure on them to undertake scholarly activity, will effectively change the incentives described above. That being the case, with new kinds of scholarship empowered and recognized, extension faculty influence on the research agenda should increase. From this writer's perspective, these potential changes on behalf of both extension and research are very positive.

The process of cultural change is indeed difficult and perhaps the integration of field faculty is even more trying than is the acceptance of the new definition of scholarship among campus faculty. There is no question that the horticultural department at OSU is valiantly struggling with the integration issues at the same time as they stretch themselves with respect to the new definition of scholarship issues. It is clear that not all campus faculty members in horticulture were happy with the integration. Some, it was said, would never accept that a field horticultural extension agent should be considered to be engaged in scholarly activity, much less participate in the evaluation of the scholarship of campus faculty. It was reported that President Byrne made clear when he announced the plans to integrate, that he expected the campus faculty would have to change more than the field faculty would have to change.

One campus-based extension faculty member serving the Sea Grant extension program as a community development specialist with a master's of science degree in the geosciences had been physically housed in the Department of Agricultural and Resource Economics for about eight years. That was a logical place for him to seek academic affiliation given that his primary colleague at the time was a full professor in that department and his work involved regional economic analysis. However, when the time came to choose an academic home for P & T purposes, he chose not to stay with agricultural and resource economics. As a practitioner of community development in a university without a community development department, he needed to find an academic home that was the closest match. The existing culture in the Department of Agricultural and Resource Economics was strongly allied to traditional notions of scholarship in that discipline and so he chose to join the Department of Political Science in the College of Liberal Arts. The political scientists like what he brings to their department and he is quite happy with their broader view of the world and interest in local community development and governance.

The most difficult integration of field faculty was the required movement of about 75 field faculty members in 4-H and home economics into the College of Home Economics and Education. The college had four academic units and about 70 faculty members divided roughly as follows:

- Apparel, Interiors, Housing, and Merchandising (AIHM)—9 faculty
- Human Development and Family Sciences (HDFS)—21 faculty
- Nutrition and Food Management (NFM)—8 faculty
- Education—32 faculty[2]

Not only were field faculty quite clear that their numbers would be threatening to the existing departments, they were uncertain that there was much of a commitment to integrate them. In the face of the uncertainty about how they would fare, the dean of Home Economics and Education came up with several alternatives for the field faculty to consider. They could integrate into the existing academic units, they could form a single, separate extension department in the college, or they could form two separate departments–one for 4-H and one for extension home economics. The field agents, with a few exceptions, chose the last alternative and formed the additional departments of:

- Department of Extension Home Economics—30 faculty
  (Since changed to Department of Family and Community Development)
- Department of 4-H Youth Development Education—45 faculty

Thus, the field faculty in 4-H and extension home economics have an academic home but they have no colleagues in their departments who are engaged in other forms of scholarship, most particularly discovery research scholarship, which can contribute directly to their programs and to whose research they can contribute. This is not to say that discovery research is preeminent or better, only that several different kinds of scholarship practiced side-by-side make each better.

Consider for example the former 4-H staff member who started an extension program in small-scale farming at the time of the integration. Because he had an M.S. in animal science and another M.A. in anthropology, he could choose to go in either direction to find his academic home. He chose anthropology and has been well received by that department. He assists in teaching some coursework and has graduate students working on problems associated with his small farm audience. The integration of the 4-H staff and the home economics field faculty into the College of Home Economics and Education was accomplished. However, the opportunity to have the field faculty influence the multitude of disciplines related to their programs and draw on those disciplines in support of those extension programs was lost. The anxiety and decision of the field faculty in the face of a potentially hostile environment is understandable, but regrettable. It's just too bad.

There are profound changes underway at Oregon State University. The climate on campus that helped to promote that change was partly stimulated by a serious financial threat to the total university in the early 1990s. In 1991, President John Byrne commissioned a Peat Marwick study that resulted in reorganization of the central administration of the university. This study also recommended that extension report to university administration. The recommendation stimulated President Byrne to ask Dr. Emery Castle to undertake a study of extension, which resulted in the Extended Education report and several recommendations that were later implemented. At about this time (1991–1993), Dr. Weiser recommended that extension administration become a university function while the extension program development become the responsibility of the colleges. Dr. Weiser also was exploring broader definitions of scholarship and the use of position descriptions in the College of Agricultural Sciences.

Castle's report led to the redefinition of the status of the leadership of extension and the integration of field faculty into academic units. Bud Weiser, while serving on the Extended Education Transition Committee chaired by Provost Roy Arnold, advocated a broader view of scholarship, basing faculty evaluation on a position description and rewarding team efforts. That committee endorsed these ideas prompting Provost Arnold to appoint a Faculty Senate Committee to review and recommend changes, if any, in the university's Promotion and Tenure Guidelines that narrowly equated scholarship with research. That Committee met regularly for a year, most of which was spent in seeking agreement on the nature of scholarship and describing it in the clearest possible terms. Once that vital concept was accomplished the details of developing recommended changes in promotion and tenure guidelines and procedures proceeded smoothly and rapidly (Weiser 1999a).

Central in that process was the influence of Dr. Michael Oriard, a professor of English and Faculty Senate President who chaired the Committee. His uncanny ability to distill the essence out of prolonged meetings and faculty forums and put it succinctly into written words proved invaluable. The Faculty Senate unanimously approved the Committee recommendations. After one year of implementation of the new model for extension, and six months into implementation of the new university P & T guidelines, John Byrne retired as president of Oregon State University. Dr. Byrne was replaced by a new president, Dr. Paul Risser, who appears to have had the good sense to recognize that something good was happening on the previous watch and has kept it going. The Provost throughout the process, Roy Arnold, was key in making the details happen. It was he who chaired the Extended Education Transition Committee that implemented the decisions based on the Castle recommendations, and he who appointed the Faculty Senate Committee to review the P & T process.

When one talks about these matters with people at OSU there is a great deal of pride in what has been accomplished and a great deal of giving credit to multiple actors. Leslie Burns said they knew that they were undertaking things that would be tough for some people but decided to work first with the folks who wanted change rather than hammering on those who were more skeptical.

The task of being the first dean of extension was viewed as being very onerous in part because it would be a lightning rod for some staff anger and frustration. It is said that at the time of the job search, the senior administrators of the university predicted that the first person in the position would not likely survive more than a couple of years. That they were so candid and forthright is remarkable. Dr. Lyla Houglum started her job as dean of extension on January 1, 1995, and is still at it. The supportive behavior of Deans of Agriculture Weiser and Dutson in the move of the extension program to university-wide responsibilities has already been mentioned as unique. Oregon State University is a very civil place to be and to visit. That would certainly have contributed to the accomplishments that have been achieved.

Dr. Weiser's retrospective on the cultural change at OSU reads this way:

> ... the processes that led to change ... (are) also revealing and interesting. For example, the process ... involved faculty input in a sustained and significant way. (T)he process was led and carried out by faculty members, although administrators often started the process. The change processes were iterative, took considerable time, and involved lots of discussion and thought. The administrators who were involved characteristically provided encouragement, but were not prescriptive regarding outcome. They were generally trusted . . . . (T)his broader vision of scholarship provided the conceptual foundation for changes that were subsequently made in faculty evaluation and the tenure and promotion process.
>
> Several other universities have tried but failed thus far in attempts to make these types of cultural change. Failed attempts that we know about started by first attempting to develop new evaluation criteria and promotion and tenure guidelines without first agreeing about the nature of scholarship. That backward approach seems doomed to failure (Weiser 1999b).

Weiser is very quick to make clear that there are significant positive efforts similar to those at Oregon State under way at both land-grant and nonland-grant universities, and that, notwithstanding the difficulties of implementing them, new definitions of scholarship are ideas whose time has come. It's happening, Weiser (1999a) says, at the university level at Iowa State University and the University of Idaho among the land-grant universities, and at the college level at other land-grant universities including Texas A&M, Ohio State University, and the University of Hawaii. Similarly, several nonland-grant universities like Portland State University, OR, Kent State University, OH, and Monteclaire State University, NJ, have implemented significant changes.

The cultural change associated with the changed definition of scholarship and its implications in other aspects of academic life is likely an absolute prerequisite—a sine quo non—for the engagement of universities with American society into the 21st century. So far as this writer is aware, notwithstanding a number of attempts underway at other universities to accomplish the change, it is only truly operational at Oregon State University. It is so tough to accomplish. While all of the implications of the changes at Oregon State are not yet apparent, it seems clear that the changes are sufficiently institutionalized that they will not easily be repealed, and are surely worthy of careful observation in the future. The changes seem promising and perhaps a real possibility for significant progress towards an engaged university with extension playing a significant role on the way to that engagement. One can also wonder if the graduate students now earning advanced degrees at Oregon State University will carry elements of that new culture with them when they become academics themselves.

One caveat about the wonderful things happening at Oregon State University needs mentioning. Notwithstanding the changes in campus and extension culture, the extension portfolio of programs at OSU still remains relatively narrow and quite traditional. The failure of the 4-H and home economics field staff to integrate into established academic departments suggests that may continue. Campus and extension rewards are not the only issues to be addressed in the broadening of the university outreach portfolio. It will be interesting to observe whether a broadening of the extension portfolio and the centrality of extension in Oregon State University's outreach activities follow as a normal consequence of the changes described above.

Wisconsin Community Resource Development: Proving It Can Be Done

Cooperative Extension in Wisconsin has always had one of the most aggressive and strongly supported programs in community resource development/community development/rural development. The Community Resource Development (CRD) designation was, for much of the past 25 years, the federal extension designation but other names are used for the function. At last count, in 1992, the proportion of resources committed across the nation to CRD programming was 7 percent.

The reason that the Wisconsin CRD program is included in this chapter as a promise and possibility is because in Wisconsin, in 1998, CRD accounted for 22 percent of the extension budget. In 1998, Wisconsin extension had Community, Natural Resource and Economic Development (CNRED) agents in 65 of the 72 counties of the state. The CNRED program was only exceeded in resources by the agricultural program, which had 27 percent of the Cooperative Extension pie. Further, by all accounts over the years, the CNRED program elicits as much or more support for extension in the state of Wisconsin than does the agricultural program.[3]

The reason usually given for failing to broaden the extension portfolio is that "no other program's supporters besides agricultural audiences step up and provide the necessary support for extension." The creed of the mountaineer with respect to political support is undeniable and hard to refute. "Don't leave go of the hold you have a hold of, until you have a hold of something else" (Webb 1997). In the struggles to broaden extension's portfolio, no sensible person has ever suggested abandoning support to agricultural programs, or not asking for support from agricultural audiences. However, the size of Wisconsin's extension CNRED program and the overall Wisconsin extension portfolio looks like they have more strong hand-holds on the mountain face than does Cooperative Extension in the rest of the nation.

A brief history of the Wisconsin program in the words of Dr. Glen Pulver, who is one of several key actors substantially responsible for shepherding the program into being, is instructive:

> In the late 1950s, a number of county agricultural extension agents in Northern Wisconsin were asked by members of their County Agriculture Committees to begin working on local issues not directly related to agriculture, 4-H, or home economics. (The counties included Ashland, Bayfield, Washburn, Iron, Oneida, Forest, Florence, Sawyer, Rusk, and others.) These committees were legislatively mandated and composed of members of elected county boards. Requests for nonagricultural educational work came to individual agents and were quite different in nature. For example, one agent was asked to begin working on economic development—largely interpreted then as industrial development. Another was asked to help administer the county forest. Still another was asked to work with resort owners on improving the tourist industry. Others were asked to work on objectives such as reducing pollution in the lakes in the counties. The County Agricultural Committees almost universally noted that there were very few farms left in their counties and expressed their confidence that the agents could make a significant impact on other county problems through their educational programs.
>
> For the most part, the agents involved were highly respected veterans of extension work, trained in agriculture, and accustomed to listening closely to the concerns of the citizens of their counties and their County Agricultural Committees. They were closely linked to specialists on the University of Wisconsin campus in Madison and skilled at gathering and disseminating useful information to the people in their counties. They were well aware of the decline in farm numbers and had already become involved in small ways in educational activities related to nonfarm issues. When this group of agents became collectively aware that they were all being asked to do nonagricultural extension type work, they sought approval for this work from the Wisconsin Director of Cooperative Extension, Henry Ahlgren. Director Ahlgren listened carefully and immediately visited with a number of the County Agricultural Extension Committee

members. He soon gave the agents and county committees his strong endorsement of extension work on nonagricultural issues with the clear understanding that the major emphasis of this work must continue to be educational.

The agents then launched substantial educational efforts directed at community natural resource and economic development problems unique to their specific county conditions. These programs were very well received locally and were soon the major focus of the agents' annual plans of work. The agents continued to serve the needs of their agricultural clients, often times sharing group and individual meetings across county lines, but the overall Wisconsin Cooperative Extension program would never be the same. Within a short time period, this group of agents sought the permission of their County Agriculture Committees and Director Ahlgren to change their official titles to County Resource Agents. Approval was quickly granted.

Shortly after the first discussions about the nature of extension work in the North, still in the late 1950s, the University of Wisconsin received a grant from the Ford Foundation to establish an experiment in urban extension. An initial effort was the employment of a county extension staff member in Columbia County with a primary responsibility to develop a program which responded to the needs of nonfarm residents in that county. The agent was trained in business, not agriculture. His early work was on broader issues of community development and included specific educational programs in business management, economic development, and tourism. Once again it was well received locally, but resulted in no further funding from Ford.

In the early 1960s, the work of the resource agents in the North and of the agent in Columbia County began to have a major impact on the Wisconsin Cooperative Extension program. Neighboring County Agriculture Committee members and others became aware of the positive consequences of the new nonagricultural extension programming and began to inquire about acquiring county resource agents for their counties. Many of these counties still had large agricultural sectors, but saw the number of their farms declining. They recognized: the growing need for off-farm employment opportunities; a decline in water quality in local lakes and streams; stresses on land use from rapid urbanization; expansion in the demand for rural home sites; and the increased complexity of local government finance. They saw that these problems were being effectively addressed through the educational efforts of county resource agents and wanted similar help. But, at the same time, they did not want to give up their county agricultural agent positions, feeling that they needed the full-time services of the ag. (agricultural) agents.

The demand for more county resource agents was intense. Once again, Henry Ahlgren listened, and with the help of his Cooperative Extension leadership team, responded positively to County Agriculture Committee requests. With critical support of Wisconsin state government and the University of Wisconsin, monies were provided to those counties who requested additional county resource agents and were prepared to provide the county financial match. As a consequence of the continued support of a number of University of Wisconsin extension administrators, the number of county resource agents grew rapidly throughout the '60s, '70s, and '80s, although, like most other public efforts, state financing became more difficult in later years. As a consequence of the growing local interest and support of the work of the county resource agents, the Wisconsin State Legislature broadened the title of the County Agriculture Committees to County Agriculture and Extension Committees.

The growth in local extension educational programming on nonagricultural issues spurred a demand for statewide extension specialist assistance in new subject matter areas. Help was sought in economic development, business management, regional planning, public finance, water resources management, land use management and planning, environmental protection, and other disciplines not always found in colleges of agriculture. Once again, the administration of Cooperative Extension and the broader University of Wisconsin-Extension were successful in acquiring needed financial support. Several existing extension specialists changed the emphasis of their work in response to agent requests. Additional extension specialists were hired to meet unfulfilled requests. Some of the additional staff were in the UW-Madison College of Agriculture and Life Sciences, some in other colleges of the UW-Madison, some on other campuses of the University of Wisconsin System and some solely in UW-Extension.

As this extension effort proved to be highly successful, concerns were raised by some state agencies. The Wisconsin Department of Local Affairs and Development, a state agency responsible for economic development and tourism began to raise questions about the appropriate bureaucratic location of the county extension staff members carrying on educational programs addressing these issues. The Wisconsin Department of Natural Resources was likewise concerned regarding natural resource focused work. The County Agriculture Committees were not about to give up these agent positions and this work to state agencies. The agents themselves indicated that their titles often confused local citizens. After lengthy discussion with representatives of the state agencies, the County Agricultural Committees, and the agents themselves, it was agreed that Wisconsin Cooperative Extension would retain responsibility for the agents, greater emphasis would be placed on their educational role, and their titles would be changed to County Community Resource Development Agents. (Today, not all agents carry this title, some are called Community Development Agents, Community, Natural Resource, and Economic Development Agents, or other titles.)

Today there are community resource development agents in 65 of Wisconsin's 72 counties. Educational programs encompass community development, economic development, tourism, business management education, water quality, watershed development, land use planning and management, environmental protection, local government administration, public finance, housing, and other issues identified by the community. County agents are hired with educational background and experience in business, economic development, regional planning, natural resources or community development. Most have their M.S. degree before being hired. The Wisconsin Cooperative Extension and the counties are joint employers of all county agents. When an agent moves on or retires, the County Agricultural Committee is expected to identify its most serious current concerns in the program area. New hires are sought with education and experience related to those concerns. For example, a county with a serious water quality problem would seek out a new agent with natural resource education and experience. Later, intense in-service education is provided new agents in those subject matter areas where their background is weakest.

The Wisconsin extension effort is currently the largest extension community resource development program in the nation. It continues to prosper while programs in other states decline. This program has succeeded for two primary reasons: 1) It was demand driven from the start and continues to be so today. Counties asked that extension do this work. Agent positions were filled at the county level at the county's request and with county financial participation. Agents work on issues defined by the citizens of the counties not some distant agency or institution. County agents and university specialists are both seen as accessible sources of knowledge useful in solving their problems. Agents are not merely gate keepers. 2) Wisconsin Cooperative Extension administrators listened to county requests and responded positively. They persisted in the fight for the resources necessary to support this effort (Pulver 1998).

There is yet another insight to the success of the Wisconsin CRD program that is not even fully appreciated by the Wisconsin staff of the program themselves. In Wisconsin, the County Agricultural Committees that Pulver reports were later changed to County Agricultural and Extension Committees are subcommittees of the County Boards of Supervisors and their membership is restricted to County Supervisors.

Thus, when Pulver states that the growth of the program in Wisconsin was "demand driven" he is talking about "demand" from county supervisors who were reflecting their constituents concerns. Requests were from county supervisors whose political base was broad enough to get them elected as supervisors, whose longevity in office was a function of their political support and legal term limits. They were prepared collectively to back their requests with political support for extension, in their county and in the state.

In almost any county in the nation, except in Wisconsin, extension is able to stack the advisory committees with friends who will say whatever the extension staff wish them to say. It is often done, or has evolved to essentially being that way, in many places in the country. Those involved in the noble efforts throughout the country working with extension leadership/advisory committees should take a close look at the Wisconsin County Agricultural and Extension Committees. It will give insight to the importance of rules of membership, terms of office, and broad representation from the community. While no state extension program will likely have the opportunity to restructure county government to coincide with the Wisconsin example, the assurance of a broad representation of the community, assured turnover, and renewed membership that is in tune with the community is worth trying to achieve and to institutionalize. It surely beats passing membership down from generation to generation in powerful families.

One last point needs be made about the Wisconsin Extension CNRED Program. While there are numerous faculty in the University of Wisconsin system who are available to be called upon by the CNRED program, the systematic organizational support for the field staff is still considerably lower than for the agricultural program. Of 197 FTE faculty/staff in Wisconsin agricultural extension supported by University Extension in 1998, 39 percent are in field assignments and the remaining 61 percent are in support positions, mostly in agricultural science departments on the Madison campus. The comparable figures for the CNRED program are a total of 141 FTE faculty/staff of which 48 percent are in field positions and 52 percent in support positions. Many field educators in extension will debate whether they get their money's worth out of campus-based staff in support of their programs. However, the argument about engagement and knowledge-based programming and the evidence of high returns to investment in agricultural research and extension suggests that the investments on the campus in support of programs is very important.

Parents Forever—A Program for Kids of Divorce and Their Parents

It's a good thing the Roses never had children. The not-so-very-funny dark comedy film, *The War of the Roses,* starring Kathleen Turner and Michael Douglas, portrayed what can happen when a divorce (without children) runs amok. In *The War of the Roses,* the audience is finally put out of their misery when the couple both end up dying as a result of their avarice, animosity for each other, and lack of self-restraint in dissolving their marriage.

Divorces involving children are infinitely more complex, and frequently the children are the casualties of the parents' unconsidered, unrestrained behavior. The resulting pain and suffering by adults and children alike is a part of the contemporary American experience—if we haven't been through it ourselves, we have friends who have. More than a million new children are affected by divorces each year, according to the literature prepared by the Parents Forever extension program from the University of Minnesota.

Yes, the Cooperative Extension Service—the folks who run 4-H camps and clubs and teach home canning—have something to say about managing divorce. It is a promising possibility and demonstration that extension can address some of the most trying and fundamental issues facing people in our society. In this case, addressing the pain and suffering that goes along with the dissolution of marriage. You want an "engaged university"? Parents Forever is engagement.

The five-part educational materials include the following topics and are supported with educator guides, parents' handbooks, and video materials:

- *Impact of Divorce on Adults,*
- *Impact of Divorce on Children,*
- *Legal Issues and Role of Mediation,*
- *Money Issues in Divorce,* and
- *Pathways to a New Life*

But this is more than the development of a set of educational handbooks, however important those publications are. It is a program that is in use in 61 of Minnesota's 87 counties. It has been endorsed by the Minnesota Supreme Court, which initiated state law, passed by the Minnesota State Legislature that went into effect in 1998, requiring that parents involved in disputes involving minor children attend parent education programs. The Minnesota Supreme Court purchased Parents Forever curricula materials for all counties that chose to use it as their program in meeting the requirements of the law.

It all started in about 1993 with Phyllis Onstad, Extension Educator in Winona County. Onstad, herself experienced with divorce, along with judges, women's advocates, early childhood specialists, attorneys, and other Winona County leaders, struggled to come to grips with the negative impact of divorce on children, particularly the use of children as pawns as divorcing parents work through their anger. Collaboratively they developed a program for parents in Winona County. The program was so effective that many other counties were interested in offering a similar program to address the same need in their communities. In her efforts on behalf of extension to respond to the problem, Onstad was joined by extension colleague Madge Alberts in nearby Dodge County, to further develop the curriculum to be used statewide.

This program was clearly grass roots in its origins, and substantially developed by field faculty members with the assistance of University of

Minnesota campus faculty. University faculty often talk of knowledge-based extension programming, erroneously presuming that all knowledge is at the university. We seldom talk of research or a research agenda derived from the knowledge and experience of field faculty, much less from citizens—this is an example of such a program. When the team of extension staff was finally assembled and under full swing in the development of the materials, the project involved 17 field faculty, five campus faculty specialists, a nonextension faculty member in human ecology, three administrators, nine communications and educational technology staff members, and six outside consultants. The program development group, to their great credit, obviously overcame the campus/field disconnect discussed in Chapter 7.

Concurrent with the efforts of the extension team two other significant developments occurred. The family law committees of the Minnesota House of Representatives and the Minnesota Senate requested Extension Educator Onstad and Judge Margaret Shaw Johnson, a key member of the Winona County collaborative team, to testify. They were asked for their professional perspectives on the importance of divorce education for parents, as family educator and family law judge respectively. They also reviewed for the committees the scope of the pioneering divorce education program in Winona County and the participant evaluation results, which were very positive. An outcome of their testimony was the drafting of bills in the house and the senate requiring parents involved in disputes involving minor children to attend a comprehensive divorce education program.

Onstad then took the initiative to contact Minnesota Supreme Court Justice A.M. "Sandy" Keith who had already made public statements about the need for more community support for families facing divorce transition. Judge Keith requested a meeting with extension administrators as well as Onstad and Alberts to share his concerns for Minnesota's divorcing parents and to learn more about the educational program development work underway within extension. This contact with the Supreme Court made it possible to mirror at the state level the collaboration between extension, the courts, and other community groups that had initiated the program in Winona County.

As extension program and materials development progressed, the team kept in touch with the leaders of the court system. When the law was passed specifying the character of the content for an approved instructional course for divorcing parents, it looked a lot like the table of contents for the Parents Forever program. The Parents Forever curriculum materials were released in September 1997 and the law went into effect on January 1, 1998.

Thus far, the program is reaching approximately 3,000 parents a year within Minnesota, and has been, or is being, adopted in five or more other states. Minnell Tralle, State Coordinator for the Parents Forever program, says that the Parents Forever team at the University of Minnesota also has had conversations with significant Hispanic community groups who will collaborate with extension in getting the materials translated to serve the Spanish-speaking community.

Parents are reporting significant progress in keeping their children out of the middle of their disagreements in the divorce, in putting the best interests of the children first, and in permitting children access to both parents. Other evaluation and impact information suggests that the course is helping parents understand the importance of child support and the real costs of raising children. In this context, it is widely acknowledged that improved compliance with child support obligations provides benefits to children, families, communities, and governments. Evidence is also clear that there is a cyclical relationship between child support and visitation. "Parents who visit pay child support and those who pay child support visit their children more—two aspects vital to the well-being of children" (Minnesota Extension Service 1999).

The program has brought Minnesota extension to the attention of entirely different collaborators than ever before, namely, the court system and the legal community, particularly those dealing with family law, and mental health professionals among others, at both the state and county level.

This program on the problems facing divorcing parents provides assistance to people from all segments of the society. It cuts across income, occupation and profession, ethnicity, and location of living. In that regard, it is unlike many extension programs that are frequently based on aspects of what people do for a living, where they live, or what they do with their spare time. The Parents Forever program is valid in any county in the state (or the country) and provides an immediate basis for collaboration in any county that chooses to use the program. It also gains strength from the presence of extension offices and extension educators in each of the counties. It was started by the insights and concern of field faculty who worked collaboratively with groups at the local level and then broadened the program by bringing campus faculty and outside consultants into the development process. As extension seeks to broaden its program portfolio and engage new and different audiences, the example of Parents Forever is worthy of emulating in other areas of people's lives—after your state's extension program has purchased and incorporated this particular program into its own extension portfolio.

Serving All Kids: A Promising Direction in 4-H

I pledge my HEAD to clearer thinking, my HEART to greater loyalty, my HANDS to larger service, my HEALTH to better living, for my club, my community, my country, and my world.

Thus reads the 4-H creed or pledge. Hundreds of thousands, perhaps even millions, of America's youth have been challenged by this motto over the 85 years of 4-H's life as extension's youth program. Criticism of the 4-H program has argued that it was too aggie—that the four H's stood for Hoofs, Horns, Hair, and Hide, and that it was a safe program for safe kids ignoring those who are "at risk." There was also a first rate nonaggie program in Massachusetts for urban and suburban kids called Pocket Pets—snakes, gerbils, mice, frogs, and other creatures that kids like to keep in their pockets. But there was always a kind of distortion regarding the intellectual investment in 4-H programming. We always knew more about the calves and other animals than we did about the kids.

It is curious that the 4-H programs are almost always isolated away from academic departments in a "state 4-H office." This is true, notwithstanding the almost religious fervor with which deans of agriculture and directors of extension argued that the strength of extension programming was from the integration of research and extension in academic departments. In most states, the state 4-H office houses an array of "specialists" or state-level staff. They share in either program management cum administrative responsibilities such as the State 4-H Foundation, 4-H Club Management, or managing the state 4-H camps, or subject matter responsibilities such as teen curriculum development or some other special support activity to the overall program. Sometimes there were specialists in the technical agricultural departments like animal science who had 4-H support responsibilities.

Seldom has there ever been the investment of research resources to examine the problems faced by youth generally, rural youth as distinct from urban/suburban youth, or youth in any particular community, and to then build programs around that type of a knowledge base. There was research on crop problems, animal problems, food safety and nutrition, farm management and finance, but we presumed that what we already knew would be good enough for the kids. In 1998, Wisconsin's 4-H and youth development program had 101.76 staff/facuty FTEs supported by extension. Of that total, 77.18 FTEs or 75 percent were in field assignments with the 25 percent balance in support roles. This is in comparison to 39 percent of the agricultural extension program in field positions with 61 percent in support positions.

Just as the agricultural extension program is multidisciplinary in its nature, most of the other extension areas are similarly multidisciplinary and cannot rely on a single academic department to provide intellectual support. This is certainly true of community resource development/rural development/economic development, family and consumer science programs, as well as the array of programs organized under the Sea Grant efforts in many states. However, 4-H is the oldest of such programs, perhaps the most successful, and also the most isolated. If it were not that the programs for the kids who are served are very good, and that kid's programs themselves evoke so much political support for extension generally, it is unlikely that 4-H would have survived to this day.

Dr. Dale Blyth, director of the Minnesota 4-H and Youth Development Center, argues that some of the marginalization of 4-H came about because its symbols—the four leaf clover, among others—became the label for all of extension's youth development work. Those symbols were strongly associated with the very agricultural nature of the early program (Blyth 1999). The "second class" nature of the 4-H program within the culture of extension is attested to by the fact that in many states for many years anyone seeking to be a county agent was obliged to serve (as in "do time") in the role of county 4-H agent prior to being given a job in their chosen professional area. There were no unique qualifications for being a 4-H agent, only for being an agent in another area. Fortunately that has virtually disappeared, and 4-H staff are now hired for their qualifications to do that job. However, in the Texas Agricultural Extension Service, one of the largest in the nation, the practice still exists.

This "bottom of the totem pole" position of 4-H within extension is emblematic of children at the bottom in society generally, not making economic contributions and requiring a lot of work that is not valued because it is unpaid. This attitude, along with the separation from supporting academic departments, appears to be reason for some of the criticisms directed at 4-H. It is generally agreed that 4-H is great for those it serves, but that it is serving a small segment of young people in the community. It is still heavily aggie-oriented, and for the most part it knows more about the animals it teaches the kids to raise than it does about the kids. Dr. Blyth (1999) says that traditional 4-H programs are the best examples of youth programming that have failed to take account of the advances in knowledge of adolescent development, whose science has come into its own since the 1980s.

But there are some changes afoot in the land-grant universities' programming for youth. Those changes reflect the understandings embodied in the now famous book title by Hillary Clinton (1996), *It Takes a Village: And Other Lessons Children Teach Us*, which suggests that child raising is the responsibility of the whole community. The 4-H/Youth Development Center at the University of Minnesota represents such a change, since it has assembled a faculty of researchers with backgrounds appropriate to doing research as well as to programming on behalf of youth. Increasingly 4-H county professionals are handing over the management of 4-H clubs to either volunteers who assist the clubs or to extension assistants who mainly handle the clerical/organizational function necessary to keep the clubs going. This relieves agents of work that must be done, though not necessarily by them, and allows them to engage in work in the community dealing with youth development problems or programs. In some places where this is happening and where the traditional program is strong and demanding, there is a second 4-H agent who does the youth development work. Such arrangements are highly complementary and suggest that extension programming for youth can serve all kids in the community.

A major effort in 4-H in most states is to bring 4-H materials, if not clubs, into the public schools. This assists in the coverage and numbers game that revolves around the Federal Form ES 237 by which counties and states report the number of "unduplicative contacts" between extension programs and children. In Minnesota in 1998, the number of kids in clubs was about 32,000, and the number of total kids contacted was around 320,000.

More significant than simply making 4-H materials on incubating chicks, or some other such curriculum available in the classrooms of the schools in the community, is the involvement of 4-H staff as true Youth Development Specialists. Some of the very best of this work was started by field staff in home economics/family development and community development as well as by 4-H staff. In a number of places, field staff have actively participated to help efforts to assess the circumstances of all of the youth in the community, rather than just doing safe programs for safe kids. In New York state where under state law a County Youth Commission deals with issues of youth, much of the Youth Commission work in many counties is carried out under contracts or grants to the local extension office (Bonaparte-Krogh 1999).

One of the very best examples of such youth development programming has resulted in some county governments requesting that extension serve the community by providing a youth development specialist. Some of the tools of analysis that have contributed to this move is some work carried out in Wisconsin by a couple of development psychologists from the University of Wisconsin-Madison by the names of Riley and Small. David Riley's scholarly interests are primarily with young children, mostly preadolescent, and Steven Small deals primarily with adolescents. Riley was frustrated. The extension educational programming that he prepared and distributed about latch key children was substantially ignored as being irrelevant to the kids in the communities where he was making it available either as a speaker or in published form. He was challenged to find out what really were the local circumstances and behaviors of latch key children. With the collaboration of some human development/home economics field educators, Riley proceeded to survey local communities on questions about the circumstances of unsupervised young school-aged children in the community.

Riley and his county extension collaborators discovered that local people were much more attentive to data that was specific to their communities, even if it totally mirrored other national research results. This was particularly true when people in the community had been active participants in the data collection and manipulation. Over the course of several years in the late 1980s and early 1990s, Riley carried out similar surveys in more than 100 Wisconsin communities. He did this despite the consternation of his academic colleagues on the University of Wisconsin campus and elsewhere in his profession who thought he was committing academic suicide (Riley 1999).

The results of these community-based research studies led to the establishment of at least 96 new school-age childcare programs, creating 406 jobs, and caring for 6,754 children. The project led by Riley averaged one business start (or expansion) every month for more than seven years. "But more importantly, each community with a new school-age childcare program was a fundamentally different place in which to raise children. The community ecology had shifted in a way that helped families, and the change took root" (the programs are largely still operating today) (Riley 1999a).

Following on Riley's experience, Stephen Small developed the Teen Assessment Project (TAP) addressing issues related to preteen and teenage youth. During the period 1989 to 1994 more than 60,000 students in 175 school districts in 40 Wisconsin counties completed TAP surveys that asked "questions on issues such as sexuality, alcohol and other drug abuse (AODA), mental health, interactions with peers, family relations, perceptions of school and community, and future aspirations" (Lande 1994).

Fundamentally, these efforts by Riley and Small prove that the same model of engagement between campus-based scholarship and field application, including the test of workability, which has been so productive solving farmers problems in the agricultural sciences can also be applied to the problems faced by children and young people. As a scholar, Small (1998) reports that prior to his work in the communities he had no particular scholarly interest in adolescent sexuality and sexual behavior issues. As a result of his findings in communities through his extension programming, he has published several articles on the subject in scholarly journals and is increasingly known as something of an expert in that area of scholarship.

Engagement makes the scholarship better. The separation of extension youth programming from the supporting disciplines is intellectually bankrupt and should be abolished.

The culmination of the youth development experience is in Wisconsin in the form of requests from several Wisconsin counties for youth development specialists who deal with the problems of all the county's youth. Part of the change clearly involves a question of the continued relevance of traditional 4-H programming to all kids in the community, particularly to kids in urban settings. Another part is a belief that extension has the talent and access to skills and knowledge that would make it relevant to solving youth development problems generally. Tom Riese, who served as the 4-H and youth agent in Waukesha County, Wisconsin, for more than 20 years is now the Youth and Family Development Educator in that county. A second 4-H agent continues to support the strong traditional 4-H program in the county. The job Riese now holds started in July 1, 1994. Riese (1999) reports that 4-H was not included in his title at the request of the county leadership, not because there was an aversion to 4-H but because they wanted it clear that Tom's responsibility was much broader than the images conjured by the 4-H symbols. This is the same argument made by Dr. Blyth and reported earlier.

Consider a few of the activities that Riese is engaged in as part of his job. They speak reams about the promises and possibilities of this type of youth programming in extension:

- Understanding Your Role as You Parent Your Teenager—workshop for parents of teenagers;
- Fathers' Night Out—and other father education programs;
- A Community That Cares—community action program that has led to applications for Community Development Block Grants, etc.; and
- Muskego: A Community for Youth—work with all agencies already serving youth of the community to engage both youth and adults in community decision making, building community spirit and pride, and to establish policies for healthy youth.

Increasingly, other 4-H agents and human development agents are moving into that more interventionist cum community development mode of operating with respect to youth development. The work of Riley, Small, and the Center for 4-H and Youth Development in

Minnesota will be important resources for them. It's long past time that we knew at least as much about the kids as the calves, and that extension had something to say about the present and the future for all the kids in the community.

Reporting Tough Issues—Oregon Extension Communications Staff as Journalists

There is within the extension system nationwide a predisposition to providing answers to people's problems based on the very best science available. When stated this way, the previous sentence reflects the noble values of extension people and most academics who wish to have their work be useful and based on the best knowledge available. However, not all problems of the society are amenable to the expert model of extension education that is mostly reflected in agricultural extension programs. More simply put, many issues in contemporary American society confound the most diligent of scientists and extension educators because they are problems about which the experts disagree. For such problems, there is frequently a great deal of science involved but some of it seems contradictory and much is simply not known or understood. They are often controversial issues. They are issues over which people fight and they frequently involve public policy.

In a number of settings, particularly local settings, the approaches of "engagement," "community-based research," or even "community development" offer participatory approaches to solving some controversial problems. But there are other problems whose dimensions are so large, involve so many different viewpoints, or whose resolution appear to disproportionately harm some interests and help others, that they seem to elude extension's capacity to speak to them in any meaningful way. Frequently the "table" at which decisions are made are in state houses or the nation's capital, and so "bringing all the parties to the table" as in the participatory approaches to engaged scholarship is not possible. Nonetheless, citizens have views on these matters, and need and wish to be better informed. Salmon in Oregon is such an issue and so is poverty in Oregon.

There are literally dozens of perspectives on the future of the salmon in Oregon—on whom is responsible for threats to it, on what the best approaches should be to salmon restoration, and on who should change their behavior. There is also disagreement about what the best science and knowledge has to say about the consequences of various causes of the problem and approaches to solutions. They are positions over which people will fight.

Enter Oregon State University Extension and a very clever technique for public education on controversial and complex issues. A production team of six extension communications staff—the folks in extension upon whom most extension educators rely to help smooth language and correct grammar, provide appropriate photos, and otherwise get publications in shape before printing—were set to the task of addressing the salmon question as though they were newspaper reporters. They set about to prepare a tabloid-size supplement to be printed on newsprint for insertion into Oregon daily newspapers. They titled the 24-page piece *A Snapshot of Salmon in Oregon*.

Except for a brief introductory statement authored by Dr. Paul Risser, president of Oregon State University, and Dr. Lyla Houglum, dean and director of OSU extension, all of the articles in the publication were authored by one or another of the extension communications team of "reporters." The articles report on research and opinions of 51 different experts, mostly Oregon State University scholars, in subjects from history to fish biology to rangeland resources to horticulture to forestry and economics.

The first section is a primer on the biology of salmon. That is followed by sections detailing influences on the salmon that include mining, forestry, ranching, farming, dams, urban life, hatcheries, commercial fishing, recreational fishing, the native American fishery, estuaries, predators, natural fluctuations, and the ocean. The tabloid ends with a summary section called "Cumulative Effects" that then describes several of the efforts at salmon restoration. Finally, there is a discussion about where to find additional information.

Some 650,000 copies of the salmon tabloid were produced and primarily distributed through insertion in 11 newspapers, including the state's largest daily—*The Portland Oregonian*—in September 1998. The publication had 17 separate reviewers listed by name and affiliation. The affiliated organizations range from The Columbia River Alliance to Oregon Trout and the Confederated Tribes of the Umatilla Reservation.

According to Dr. Peter Bloome, assistant director of OSU extension, who brought the technique to Oregon with him from Illinois, the tabloid's approach generally follows the definition-alternatives-consequences model for public policy education. Bloome goes on to describe the process involved in developing the tabloids:

> Early in a project, an advisory team is created. It comprises both faculty and external advisors. I want to have some advisors with clout. For (the poverty tabloid) we have the head of the Oregon Progress Board and two of the governor's aides. This lends credibility and helps keep the agencies

from being too concerned about the project. The advisory team meets with the production team to talk through the issue, help the team develop an outline, and suggest resource people.

The production team produces the draft articles and does source checks on all quotes. The advisory team and the production team suggest individuals for the review team and the review session is scheduled (Bloome 1999).

According to Dr. Bloome the review process is critical and partially determined by the topic. For example, the salmon tabloid was more controversial so the review process was more tightly controlled than was the case for the poverty tabloid. In the poverty work, the greater challenge was to portray how the several poverty factors play out in different ethnic groups without making poverty an ethnic issue—avoiding the flawed culture or flawed character explanation of poverty.

Reviewers are invited to come to Oregon State after all of the articles are drafted. They are asked to review the entire publication at a single sitting and to then decide whether they would individually permit their name and affiliation to appear on the publication. To make that decision, individual reviewers had to evaluate whether there was a fair treatment of their point of view and whether it was presented in context that was fair to their view. According to Dr. Bloome, there were no reviewers who declined to allow their names and affiliations to appear on either publication.

> I set the context for the review session. Only the author can change an article. Words can not be placed in or taken out of anyone's mouth. All perspectives are to be included. I ask the reviewers to help us ensure that all perspectives are accurately and adequately included. They identify any "poison word" or concepts that might cause someone to stop reading. In general I ask them to help us improve the educational impact of the publication.
>
> They read all the articles, marking areas for discussion and making suggestions on the manuscripts. This takes about three hours. Over lunch, we begin discussion of the articles in the order of their proposed inclusion. I have to push pretty hard to move us along through the articles by the end of the day.
>
> The production team decides what to do with what they heard, individually and collectively. They have the notes the reviewers made on their articles, and they follow up with reviewers and additional suggested resource people.
>
> The interaction within the review session is rich and vitally helpful to the writers. Often the reviewers try to do the writers' work by suggesting changes in wording. The writers tolerate this, and I move the discussion along.

The production team was not happy with where they were after the first review session on the poverty tabloid articles. They asked to be able to start over and have another session. I was happy to comply even though it set the schedule back. At the second session, the articles were much improved (Bloome 1999).

Oregon State University Extension has published 850,000 copies of the tabloid, *A Portrait of Poverty in Oregon,* in a state of 3.3 million people. There were 48 different sources quoted in the tabloid, some from OSU but many from outside the university. The tabloid was inserted in 20 newspapers around the state in January 2000.

The very special character of this approach to informing citizens about controversial issues is that it permits extension organizations and their land-grant university to participate as a source of information about the issue without taking a position on one side or another of the issue. It is certainly in the finest tradition of the land-grant universities, which were established to make American democracy better, as was argued in Chapter 1. Further, it advances the public image of the university and of extension as a source of the best knowledge available even when there are disagreements between the scholars about the implications of the different types of knowledge. Finally, it informs citizens that for some decisions there is not a scientific answer, but rather a responsibility for each citizen to form her/his own opinion based on the information available and their own values.

> How did we choose this (poverty) topic? Bruce Weber (OSU extension economist) suggested that our legislature would be looking at welfare reform during the intersession of our biennial legislature. We decided that welfare reform was too narrow a topic and broadened it to poverty. Since we did the first tabloid on a primarily natural science issue, we wanted to choose a primarily social science issue. Just as the salmon tabloid repositioned OSU and OSU extension in the salmon and natural resource issue, we expect the poverty tabloid to reposition OSU and OSU extension on the poverty and social services arena.
>
> I see the tabloid as an important marketing effort. It is a tool allowing us to position ourselves for the future—to cause people to see us and think of us differently. It puts our future in our own hands. It also fully utilizes the talent and skills of our professional communicators (Bloome 1999).

Certainly many extension staff, by virtue of the character of their jobs and the problems they work on, are heavily vested in the technology transfer or expert model of outreach/extension—they are used to giving answers. It is not a stretch to imagine that the tabloid technique described above will be a useful educational example and model for those faculty and extension educators who have not had experience in dealing with controversial issues. It will help them to discover that the expert model is not the only appropriate model for extension to employ.

One last comment is appropriate based on Dr. Bloome's comments about the tabloids. In recent years, in response to extension being described as the "best kept secret around," many states have invested in public relations approaches to the marketing of extension. There are banners and signs and name tags and letterhead and all manner of public relations materials being used to increase the visibility of extension. There is a great deal to be said for making sure that extension in the county, the state and nationally gets credit for what it does. The best extension PR by far is the distribution of substantive information that clearly accredits extension and its university partner/patron. The public policy issue tabloids do that in spades. Dollar for dollar they are infinitely better than spending resources on making sure the name tag extension staff wear is big enough or is shaped like the state.

Try the tabloids. They work.

Sparks by the River—The St. Louis Storytelling Festival

It was 1979. During a mid-summer lunch break, Ron Turner, associate dean of the College of Arts and Sciences at the University of Missouri-St. Louis, was ensconced in the *Wall Street Journal.* He spotted a small article buried deep in the paper telling about the New York Storytelling Festival sponsored by the New York State Committee for the Humanities. According to Turner (1980), he read and reread the account and thought, "Why don't we do something like that in St. Louis?" He tucked the idea away in his head for future reference and pulled it out in September when he was invited to submit a proposal for special projects to a special funding initiative of the university. He proposed a St. Louis Storytelling Festival. The proposal was funded.

Turner had shared the idea of a storytelling festival with some others around the university such that when in early October 1979 a news article on storytelling appeared in the *St. Louis Post Dispatch,* friends of Ron clipped it and shared it with him. The article was about how storytelling was being used in the Kirkwood elementary and middle schools in St. Louis County to teach language skills. The Kirkwood storytelling project enabled a couple of the teachers to spend full time in classes with students hearing, telling, writing, and sharing stories as part of the language arts curriculum. Dr. Turner says, "I was struck, not so much by the use of storytelling for children, but by the fact that professional storytellers were among us. It was the realization that storytelling was a profession that caught my interest. When I phoned the two Kirkwood School storytellers and asked them to tell me what they were doing, I became aware of this silent, almost underground, national/international network of storytellers who were moving among us in schools, coffeehouses, and other venues reviving the ancient art. Archibald's story was the link that I needed to become aware of that network at a time when it was gaining momentum" (Turner 1999).

In mid-October, Ron Turner convened a meeting of the two Kirkwood storytellers, faculty colleagues from the departments of speech and history, a friend at the Mark Twain Banks and a representative of the Gateway Arch Museum and suggested considering organizing a spring storytelling event in St. Louis. The group decided that the purpose would be to stimulate interest in major themes, origins, and techniques of storytelling for teachers, parents, children, and others in the St. Louis area. They also made two operating rules for the group. First, all ideas would be accepted on their merit and no ideas would be rejected during the formative stages of planning. That kept a few fragile ideas from being destroyed at the outset and some of those turned into important elements of the first festival's plan. The second operating principle was that the project should be fun for the planners, participants, and the storytellers. That continues to be the goal of the Annual St. Louis Storytelling Festival (Turner 1999).

The Storytelling Tradition

> A river of stories flows from an unknown source, through each culture, through all time, through today to tomorrow. Each year, storytellers and audiences gather for warmth by the story fire, tending and telling tales, and keeping the sparks of story aglow by the river (Kammann 1999).

If you want to know more about the St. Louis Storytelling Festival, check out the Web pages listed for the Festival and for the Jefferson National Expansion Memorial (Gateway Arch and Old Courthouse) National Park.

Every May since 1980 as many as 100 professional and amateur storytellers gather on the banks of the Mississippi to tell their stories. Thousands of people—about 24,000 people each year—school children and their teachers, young people and old people, rich and poor, deaf and hearing people, families, and individuals—come to the Jefferson National Expansion Memorial and its Gateway Arch to experience the magic of the storytelling in the premier storytelling festival in the nation. The Festival has been featured on the Voice of America, on Mutual Broadcasting's "What's Right With America," and was included in the 1986 publication, *America's Best Festivals, Celebrations and Parades*. It has been written up in *Time, Parade,* and the *Los Angeles Times*. The Festival partners with numerous schools, libraries, parks, community centers, hospitals, and detention centers, to make the storytelling accessible. "The idea of partnership and inclusiveness is something we work hard to bring about, and I think it adds to the success of the Festival," says Ron Turner (1999).

The St. Louis Storytelling Festival is an outreach program of the College of Arts and Sciences of the University of Missouri-St. Louis, and University Outreach and Extension of the University of Missouri.

Whether a program such as the St. Louis Storytelling Festival should be organized under the aegis of Cooperative Extension or some other outreach function of the university is really of little consequence, so long as it happens. As a result of changes at the University of Missouri during the 1990s, Ron Turner is, at the time of this writing, vice president for University Outreach and Extension, so it is indeed a formal part of the University of Missouri's Outreach and Extension. Mostly, this writer is thankful that Dr. Turner had the sensitivity, sensibility, and energy to make this wonderful event a repeated happening.

Conclusion

Writing this chapter has been the most exciting and pleasant part of the work on this book. It is so much more fun to say complimentary things about people's work and to elicit their input to bragging on them, than it is to write about dysfunction in the system. Nothing is more renewing of one's belief that the land-grant extension system can still function in the ways that it was supposed to function for all of the people than examples that prove it is still happening.

The first section on humanities extension at North Carolina State University and the last one about the St. Louis Storytelling Festival were particularly interesting to the author. They illustrate that extension—the engagement of universities with the society—is about moving people's minds in many different directions. Moving people's minds is much more than any single model, such as technology transfer or the provision of experts to solve problems, though scholarship is still important. In the storytelling festival, the "scholars" or artists were not, for the most part affiliated with the university, and so the festival is as much an informing of the university community as well as those attending the festival, of the artistry and scholarship afoot in the land. It is in Bok's terms the university acting in its fundamental role and obligation to civilization "to renew its culture, interpret its past, and expand our understanding of the human condition" (Bok 1990, 104). However, in the case of the story telling and in contrast to Bok's position, without the engagement of the festival, the university might not know that part of our culture.

Outreach in the humanities is also interesting because throughout the writing of the book the author has used a fictional humanities extension program as a kind of trial balloon in interviews to find out where people were with respect to a vision for extension and land-grant university engagement, but that, dear reader, is a story for the next chapter.

It may only take a few Ron Turners in history before a history extension program can be a part of the engagement of any land-grant university with its community. However, for it to happen widely and become part of the norm of what land-grant engagement means, may require that the kind of institutional change underway at Oregon State be more widely a part of all land-grant universities.

Notes

1. Dr. Miller was an extension agent in West Virginia from 1939–1942; professor and extension specialist in sociology at Michigan State University, 1947–1955; Deputy Director and Director of Extension, Michigan State University, 1955–1961; provost, Michigan State University, 1959–1961; president, West Virginia University 1962–1966; assistant secretary for Education, HEW, 1966–1968; president, Rochester (NY) Institute of Technology, 1969–1979; among other positions.

2. The integration took place in 1995 and the numbers for the listed departments are the current numbers from the college's Web page.

3. Personal conversations from early 1980s to the present with Carl O'Conner, Ayse Somersand, Ronald Shaffer, Glenn Pulver, and Patrick Boyle, all of whom had leadership roles in Wisconsin Extension or in the Wisconsin Community, Natural Resource & Economic Development Extension Program.

9

Imagining Extension in an Engaged Land-Grant University

Introduction

This chapter is not a conclusion. It is a further working of the analysis and ideas set forth in earlier chapters with a view to projecting or imagining a land-grant university and an extension system associated with it into the 21st century. The imagined institution will have what this writer believes will be required to establish a new social contract between the university and the community of people to whom it belongs. The imagined vision for the future is simply that, since all of the requirements for success are not yet in evidence in any of the land-grant universities. However, various elements of the institutional requirements for an engaged institution and an effective extension effort are present several places within the land-grant system.

As part of the research process for this book, the history extension program described below was imagined and used as a trial balloon to see what visions people had for extension. Its modus operandi is very much in line with much of traditional extension programming, and it appears to have the political support base needed, and yet is in the humanities. It's a good place to start the imagining process.

Consider an extension program in history. The program has initial funding and two faculty historians each with 50 percent appointments in extension. They may or may not earn tenure but have that opportunity because they are in an Oregon State University-like institution with its definition of scholarship and position descriptions. Their extension activities are directed at providing assistance and extension education to local historical societies. They carry out programs on historical preservation based on legal provisions and resources available to that activity. They teach, or design programs to be taught, about recording oral histories, and indeed may involve some of their undergraduate or graduate students in such projects with communities. They design and teach, or have taught, programs on small museum management and operation. They make local historical societies aware of archived materials, particularly photo archives that might be relevant to them. They teach the practice of historical research to interested members of the historical societies and assist them in the codification and publication of their work. They teach or have taught genealogical techniques.

Most every county in the nation has a historical society. In many of those societies, the membership represents some of the longest-term residents of the community and therefore some of the most influential in the community, so one can easily imagine that they will be as willing as farmers to give voice to their support of extension programming. The model, like the NC State Humanities Extension, is the classic extension model, though not necessarily heavily county extension office dependent. One can also easily imagine that the history taught in the campus classrooms and practiced by the extension historians, and even their colleagues, will be somewhat more relevant to contemporary students and perhaps even better history.

Ideas and Thoughts That Lead to Further Imagining

What Students?

The Kellogg Commission Report (1998) *Returning to Our Roots— The Engaged Institution* makes very clear that "engagement" includes involving students in community as a part of their learning experience. By "students" they mean even, or primarily, residential, undergraduate students. The point is made with such emphasis that it almost leads the reader to believe that the major point of "engagement" as set forth in the report is to improve the learning/training experience for those students normally served by formal classroom instruction in support of degree programs. It is certainly true that engagement can and should include enhancing classroom instruction and the learning experience for degree-seeking students registered with the university. However, and not really at odds with the Kellogg Commission report, the notion of who the university's "students" are, was substantially changed by the land-grant universities and their history. That is part of the original social contract and must be part of the new social contract.

The notion of engagement here set forth, and normally considered by university faculty and staff involved in outreach, is of an involvement with people who may never qualify to sit in the university classrooms as degree-seeking students. That is certainly the record of the land-grant history as illustrated by Norman Rockwell in his "Work of the County Agent," reproduced in the frontispiece and referred to in Chapter 6. The "students" of an engaged university as here envisioned may be corporate leaders in a management workshop put on by a school of business or it may be welfare mothers in an Expanded Food and Nutrition Program (EFNEP), as is now widely carried out by extension in many states. In many places, the students are still three generations of a farm family as illustrated by Rockwell, even though at least two generations of those members of the farm business have degrees and sat in classrooms in the land-grant university.

The other major emphasis of the Kellogg Commission discussion of the engaged university is that engagement is more than technical assistance, as in a one-way transfer of information from the university as the center of knowledge to the ignorant masses. The argument is important and worthy of the emphasis they give it because many academics can get a little arrogant about the role of the university in the society, a la Derek Bok as discussed earlier. Within extension, there has long been a debate about the character of the extension model. Is the extension model best characterized as technology transfer by people with answers, or is something more complex involved? The notion of engagement understood here accepts the Kellogg Commission emphasis, but distinguishes that different kinds of information and different kinds of problems elicit different kinds of responses from an engaged university.

Merrill Ewert (1999), director of Cooperative Extension at Cornell provides a useful way of thinking about various approaches to outreach activities. In a 2×2 matrix as illustrated below, with information content on one axis and process content on the other, he defines four different educational strategies.

The low content/low process type of educational strategy that he calls "service" might be represented by soil testing within the existing experience of extension. The high process/low content cell in the matrix that Ewert calls "facilitation," can be an important skill in many public policy/community development programs. However, the lack of content in the strictly process (facilitator) approach may account for some of the negative early reaction to Community Resource Development programming when some CRD staff had nothing but process skills to bring to the table.

The cell of high content/low process that Ewert calls "content transmission" is the character of the educational approach in technology transfer. It has been the dominant educational approach involved in the agricultural extension program for many years and accounts for the preoccupation of many extension agents with providing "answers" to people's problems. It is also the reason many agents are uncomfortable when faced with problems for which there is no scientific answer.

The high content/high process cell that Ewert defines as "transformational education" results in transformational learning, and Ewert

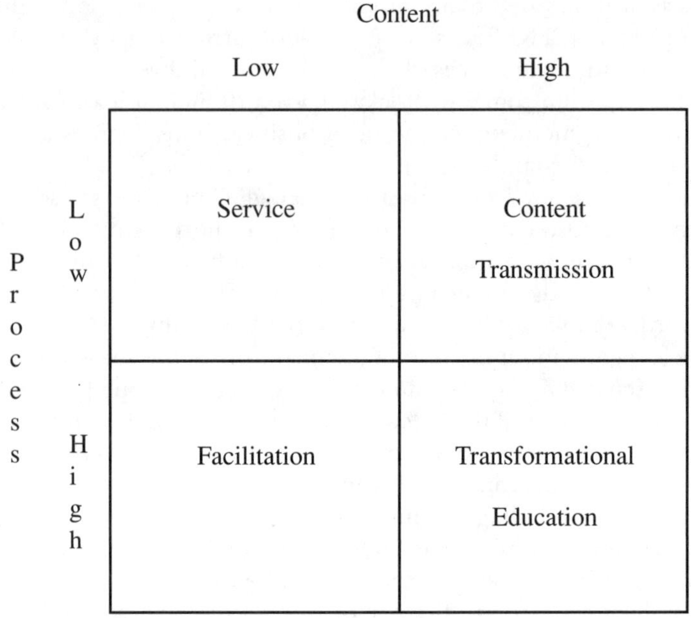

	Content	
	Low	High
Process Low	Service	Content Transmission
Process High	Facilitation	Transformational Education

Figure 9.1 Ewert instructional styles

argues that this approach to education is the most productive and transforming of people as they struggle with solving problems. It is essentially the transformational educational approach that the Kellogg Commission has in mind when they write about engagement. It is also this type of learning that drives the community-based or community-collaborative research movement. The argument is that if, as in the Riley and Small work in Wisconsin, local people participate in the research (process) they will act more responsively to the results (content) of the research. It's true—they did (see Chapter 8 for details). However, as has been pointed out elsewhere in this book, not all of the problems of the society with which the university must engage will be amenable to either the community-collaborative research approach or the "transformational learning" educational strategy.

Further, there is still validity and a role in university outreach for the other types of educational approaches in the Ewert matrix. The arguments about the contribution of the outreach to scholarship, while somewhat less visible in the approaches other than the transformational learning one, still are possible and still are important. Consider that conflict resolution is a further refinement of "facilitation" and is increasingly important in the resolution of problems in the society. An increasing number of extension faculty are picking up that skill for use in their programming. Consider also that the "training and visitation" approach to extension in agriculture, which is the approach adopted by the World Bank in its funding for Third World development, is wholly a technology transfer (content transmission) approach and in some settings has an important place. Finally, where a university has the only piece of equipment in a region that can solve a particular analytical problem, as was once the case with soil testing, or where the university has a unique resource like a botanical garden or museum, the low process/low content approach is certainly appropriate.

In the context of asking the question of this section, "What Students?" it is important to recognize that coverage is important. That is, when the land-grant university undertakes to engage the people of its state with programs to deal with particular problems, one of the criteria for the success of the engagement is coverage of the affected audience. That is, when David Riley set about to assist the parents and citizens of Wisconsin counties to understand the problems faced by latch key children, carrying out the community-collaborative research on one or five counties was not sufficient. He did it in 100 counties because that is what he needed to get the coverage necessary to deal with the problem in the state—it took doing the work in 100 counties to reach a significant number of the "students" who needed to be engaged.

What that means in terms of the Ewert matrix is that sometimes in order to achieve the coverage required, some extension/outreach programming approaches will begin to look like technology transfer even though transformational learning is what is involved and intended. For example, after carrying out a data collection procedure in support of some analysis in a few communities, the extension educator develops support and instructional materials to be used in data collection in the next 99 communities. Much of the work in the 99 communities on how to collect data may look like "content transmission." But it is the data collection that started the whole process and served as a transformational learning experience creating a desire to receive and respond to what is learned about the community.

The several educational approaches to engagement make clear that the different approaches have different strengths and advantages in different circumstances where the university is seeking to engage the society. Whether those it engages are considered "students" or "collaborators" may depend on the educational strategy employed and on the character of the knowledge being shared. However, what is clear is that many more students are involved than those registered in university degree programs—some of us who hold academic rank in the university may even become "students"—and that's the point of engagement.

Information Technology

There are many forces on the horizon that will change the way that universities do business, not the least of those is the World Wide Web and information technology. This book does not, and will not, deal much with the impacts of that technology on the role of either land-grant universities or extension into the 21st century. This omission is both because of the limits of the author's insights on such matters, and because of the belief that the fundamental incentive systems already discussed will still play themselves out in the new technological environment in much the same way as in the past. This is not an underestimation of the impacts of that technology on the way the instruction and knowledge handling ("delivery" is too limited a concept in this context) will take place within the academy and between the academy, its partners, and its public. Those changes will be profound! However, they will not solve the problems of extension and the academy identified in this book, and, if anything, will likely only exacerbate the problems already described.

For example, where there are disconnects between campus resources and extension field offices on programming, one can imagine that the gap will be even wider. The field will have available to them, through information technology, ever more and different sources of information besides the folks on their own land-grant university campus. Where campus extension people find the field faculty or staff unresponsive with respect to their programming areas, they may have greater ease of simply skipping the field staff in the delivery of their programs to a final audience using or assisted by information technology. Where there is a good working relationship between campus and the field, the information technology will make it even more productive. For any university faculty member who chooses to engage in outreach or community based research/engagement above normal duties, the information technology may very well make that involvement less costly and easier to accomplish. However, the technology is unlikely to have any influence on the culture of the scholarly associations and what is deemed appropriate scholarship and what is not so anointed.

University Engagement Must Involve All of the University

The larger challenge to the future of the university and to extension as it seeks to play a role as educator, of both the university—to the problems of people—and of people in their communities, will be the extent to which the total university is challenged and engaged. There are clearly segments of the university whose expertise will be highly valued in commercial markets. Universities will seek to capture as much of the value of that knowledge as it can through copy and patent rights, and direct or indirect commercialization, which has been encouraged by the 1980 Bayh-Dole Act, that for the first time allows universities to patent the results of federally funded research. Some in universities will call that engagement, though other forms of engagement from those fields need also to be explored.

Some of the commercialization, like the selling of the UC Berkeley, Department of Plant and Microbial Biology to Novartis Corporation, the Swiss firm, will clearly participate in defining the boundary of what is appropriate "pricing" of intellectual property and what is inappropriate (Press and Washburn 2000, 39–40). Commercialization will be particularly the case with knowledge in science and technology that leads to new marketable products. However, there will be other segments of the university whose ideas and insights may seem less directly applicable to problems of the day or may seem to be of less commercial or even extension value. As examples in the preceding chapter, "Promises and Possibilities," make clear, some of the application and realization of scholarship in the context of outreach type programming takes real creativity, and may have yet to be fully appreciated. Other fields of study embraced by the university will simply be ignored because they appear to have no or limited commercial value.

It has been argued at great length that engagement of the university in the society results in both an improvement in the society and an improvement in the scholarship of the university by virtue of the engagement. It is also likely that there will be an improved appreciation of the scholarship of the university by the society. When translated into public budget support for the university, this is substantially why university presidents are particularly interested in it as was argued in Chapter 2. In the pursuit of engagement of the land-grant universities in the society, the breadth of the engagement across the university, for the sake of both the university and the society, must be one of the standards of measurement that is applied. This is a point nowhere mentioned in the Kellogg Commission report.

Expressed in another way, Dr. Maria Tymoczko, professor of comparative literature and specialist in Celtic studies, writes about the threat of demise of the traditional humanities in the 21st century:

> From being an independent force in society, a site of discussion and debate about alternative intellectual and experiential possibilities, offering a counterweight to other social institutions, the University could become increasingly a servant of hegemonic interests, bought and sold to

the highest bidders, those with the most power and influence. At the same time, paradoxically because of their power and defining presence in the world, dominant cultures would become isolated in their ignorance of cultural alternatives.

. . . What is at stake here is the shape of academe in the 21st century. Although the first challenge to the survival of any field must be taken by professional societies and scholars themselves—namely bringing the field in phase with current intellectual practices and current views of knowledge—the second challenge cannot be met solely by the individual fields themselves. The domain of the university is the responsibility and purview to whom education and the production of knowledge are entrusted. Through the dialogue that will ensue about these issues, I believe that the nature of the university will be determined as much by epistemological, philosophical, social, and ideological factors shaping our views of knowledge as by changing technologies (Tymoczko 1999, 16–17).

The fictional history extension program described above was used in the research for this book primarily because it is an example in the humanities where the extension programming potential is not immediately obvious, and did not involve technology transfer, the traditional educational strategy of agricultural extension. It was chosen because it threatened the notion that the extension model was unique to the agricultural sciences and the farming/agricultural community, and from this point, extension was therefore most logically under the control of the deans of agriculture. The history extension program example may be even more important for its symbolism of the need to have the society understand, engage, and affirm more than the technological part of knowledge and the university. Indeed, it may be that the fullest appreciation of the university in the society will come through extension and outreach in the humanities, the arts, and the social and behavioral sciences, rather than in the physical and biological sciences.

Apropos the notion of extension based on the behavioral sciences, the engagement reported under "Youth Development" in the preceding chapter is a really solid example. Steven Small and David Riley worked through county extension staff with communities to bring the insights of child psychology and the sociology of the family into play in solving problems of latch key children and preteenagers. Their work is akin to the development of hybrid corn by agronomists for farmers. Indeed, the work may be more significant than the hybrid corn experience because as Riley (1999) reported, he was never was able to get the job done, practically or intellectually, until he engaged the communities in the research. There is a whole literature on community-based collaborations and research in the behavioral sciences and the advantages the approach offers those disciplines (Lerner and Simon 1998).

Finally, if Derek Bok's (1990) position is correct—that as universities engage the world, they must be careful that they do not "succumb to its blandishments, its distractions, its corrupting entanglements" (Bok 1990, 103–104) diminishing the "more profound obligation that every institution of learning owes to civilization to renew its culture, interpret its past, and expand our understanding of the human condition" (Bok 1990, 103)—then there is no need to be concerned about the extent of engagement across the whole university. Whatever the universities offer the society are wonderful gifts from on high and the people should be appreciative of the offering. However, if the arguments made in this book are true—that without engagement, public universities in the 21st century will not be able to understand contemporary culture, much less renew it, will not understand the past or be able to interpret it, and will have no basis for understanding the human condition—then the breadth of university engagement is supremely important to the university.

When Given a Chance, People Say They Want Much from Their University

The Wisconsin evidence on the role of democratically chosen and functioning extension advisory committees or councils with a powerful voice at the state level is very interesting and suggestive. It is, to this author, highly persuasive that in the face of some countervailing power with an interest in extension programs, agricultural interest groups can give up holding extension hostage.

Iowa has also had what, on the face of it, appeared to be an independently functioning county council structure comparable to that in Wisconsin. The members of the Iowa extension councils are elected in open county elections, along with other county officials. However, only in the past couple of years has that group of county-elected officials had any state level organization to give them a larger political voice. By all reports from Iowa, the newly formed state organization appears to have a collective view of a broader extension than is now the case and a growing state level voice to support that broader agenda (Johnson 2000).

The Wisconsin record, the emerging Iowa experience, as well as other evidence about representative county level extension councils from around the country, suggest that when communities of people are given an honest and democratic voice on what they want, they are saying they want a great deal from their university—if they can figure out a suitable way to make the engagement work.

One of the implications of this insight is that university presidents must appreciate and support their staff as they struggle to attend to the state and local politics of extension in the counties and in the legislature.

This means that senior university administrators should avoid nonsensical rules that seek to limit political contacts on behalf of the university to some limited number of senior staff members. Such rules are naïve at best, stupid at worst, and fail to understand the character of the complex of political forces that impact on the land-grant university and its extension system.

Senior university officials must understand that, if they want their university to be a "people's university," its future will be in the hands of all of the people of the university in their myriad of contacts with the people of the state. University presidents have little choice but to trust the political judgments of their faculty members on the campus and in the counties. It is clear, given the choice between leadership and management, they must chose leadership. Given that choice and having made it, they may wish from time to time to remind themselves and their staff, including the secretaries who answer the phones, of the values and operating principles upon which the university is built, so that everyone is singing the same hymn. A portion of the Glion Declaration is certainly a start in that direction:

> In its institutional life and its professional activities, the university must reaffirm that integrity is the requirement, excellence the standard, rationality the means, community the context, civility the attitude, openness the relationship, and responsibility the obligation upon which its own existence and knowledge itself depend (Glion Declaration 1999).

An Institutional/Administrative Structure Is Necessary

In an October 1999 workshop on "Best Practices in University Outreach" held at Pennsylvania State University, Penn State President Graham Spanier, one of the most articulate and ardent spokespersons for university engagement, suggested that the engaged university would be attained when there was no need for any kind of organization to facilitate the engagement since it would be part of academic culture and would simply happen in the course of normal academic practice.

This writer disagrees with President Spanier. Consider the Oregon salmon and poverty tabloid public policy educational projects and technique reported in Chapter 8. Since the character of both salmon policy and approaches to dealing with poverty in Oregon vastly exceed the insights of any single discipline, they are well beyond the scope of "community-based research" techniques, however useful such techniques may be for other types of problems and university engagement. Large public issues with multiple facets that elude "scientific" solutions but require as much knowledge and wisdom as can be amassed, require some facilitating body to bring together all of the information and to organize and generate support for the university part in the activity.

Similarly, as any faculty member with a heavy extension appointment within the current system can relate, the development and implementation of an effective extension program in any subject cannot be done effectively in an offhand way, as part of the normal academic assignment in classroom teaching and research. As important as teaching is in the university and as important as a real world project is to the student experience, the kind of engagement that brings the university face to face with the problems of its community is much more than student *practicums* or class projects; though an engaged university will have many opportunities for such student involvement in the community.

This is as true in the humanities and the social or behavioral sciences as it is in the physical or biological sciences, where technology transfer is often the engagement model. Regardless of the engagement method— technology transfer, community-based or collaborative research, or public policy analysis and education, whether at the national, state, or local levels—the engagement is demanding and time consuming. The faculty member requires support and resources to be able to carry out the activity effectively. That means that there need to be the institutional and faculty resources to make the outreach/engagement function an equal partner with on-campus teaching and research, and to ensure that it happens, even if there are no on-campus students involved.

As important as the need for funding to pay for faculty time is the institutional support that empowers faculty to get involved outside the campus. Even entrepreneurial faculty who raise funds for outreach through grants need to obtain adjustments in their other duties so they can carry out extension/outreach programs that are independent of their classroom teaching obligations but may contribute to that teaching activity. Of particular importance is the need to be relieved from on-campus teaching duties that particularly conflict with the schedules necessary for successful off-campus outreach. In some degree, it is the question of serving one group of students or another group of students. If there is a full commitment to engagement then finding the means to successfully serve both the registered campus students and the other off-campus students whose identity is, at the start of such programs, more elusive and amorphous, is absolutely necessary.

For the university that wishes to encourage engagement of its academic units with the society, one of the most critical issues is to provide some faculty in each academic unit with at least one semester per year when they have no classroom teaching duties and are fully dedicated to outreach/extension activity. Developing and maintaining the relationships that are a sine qua non of an extension/engagement program is time consuming and demanding. Notwithstanding President Spanier's comments referred to above, the kind of support here envisioned will require administrative oversight and assistance.

For outreach/engagement programs in the extension tradition, which may or may not be university-wide in authority and portfolio, the implication is clear. They may play a role in their land-grant university's engagement or they may not. If they do play a role they will need to become involved with the total university. If Cooperative Extension does not play a broad role, it may get left behind in the dust.

The change in the extension administrative function at Oregon State University is instructive. Extension administration's role changed from program development, direction (administration), financing, and evaluation to one of oversight, macrofinancing, and evaluation, leaving program development and direction, including a substantial part of personnel matters, to academic units. The new arrangement makes it possible for the extension administration to work with any academic unit that is interested and willing to carry out an extension program and provide the support and empowerment necessary to get it done within the context of the missions of the university. It appears, at least superficially, that Oregon State University Extension will be central in the continuing efforts of Oregon State University to become a more engaged university.

According to dean of extension at Oregon State, Lyla Houglum (1999a), under the new system field faculty feel much closer to the academic units than ever before. It is certainly difficult to conceive of extension field staff playing a significant role in university engagement if they only have a sense of belonging to the extension service part of the university and little or no allegiance to the academic part of the academy.

A Change in Academic Culture Is Required

When the culture of the academy understands outreach/extension/ engagement as simply fulfilling the obligations of tenure, as discussed in Chapter 2, several things will be better in an engaged university. Faculty with partial extension appointments will have their extension-related scholarship valued, and they will not have difficulty obtaining the assistance of faculty without such appointments. Indeed, faculty with extension appointments may spend considerable amounts of time providing the social infrastructure necessary to facilitate the larger contribution of their colleagues to the engagement. The extension historian may broker a relationship between a community historical society and faculty in the anthropology department or other historians to carry out an excavation of a historical site of interest to the local group. The issue is not whether at some land-grant university such an arrangement exists and whether some historian does this—it surely must happen somewhere. The issue is can it be made the ordinary outcome of scholarly practice.

The culture of the academy is far away from a cultural norm that evokes engagement as a normal outcome of scholarly practice. Further, as was made clear in the discussion of the Oregon State University institutional changes, there is a real limitation on university administrators about what part of academic culture they can actually influence. Notwithstanding all of the changes accomplished at Oregon State University, the only real leverage the OSU administration has on faculty members to comply with the new definitions of scholarship and the position descriptions is the refusal to administratively move the paperwork of any individual unless there is an approved position description.

University presidents and senior staff need to be modest in their claims about making changes and should not misunderstand administrative changes as necessarily resulting in institutional change. Administrative changes, even ones that have been in place for a very long time, can be changed as quickly as university presidents change. The experience at Virginia Tech certainly would seem to affirm the point. An "Extension Division" with university-wide authority had operated for almost 20 years and was lauded as a model for the national system. It was returned to the control of the College of Agricultural and Life Sciences by the actions of a new president under duress from the pressures of agricultural interests, at the urging of a dean of agriculture. Administrative or organizational change in the absence of significant personnel and policy change is like trying to make champagne out of ordinary wine by moving the bottles around in the wine cellar.

In the absence of rather profound cultural change, the only alternative tools for university leadership seeking to promote engagement are money and rhetoric including cajoling. In the absence of a culture that makes involvement in the society part of the academic norm, spending significant amounts of money, principally on faculty time away from other responsibilities, is about the only alternative to obtaining active faculty involvement in outreach. The experience thus purchased may, in and of itself, accomplish some of the cultural change. It is important in evaluating the character of the changes in universities toward engagement, particularly when the leadership of the institution is vocal about the direction of the institution, to ask whether the changes will be sustained after the vocal leadership has left. In the absence of significant change in the culture of the academy, having just the support of an affirming university leadership with its rhetoric and cajoling, is a lot better than nothing.

Until there is discussion within the scholarly societies about the character of the scholarship they are prepared to affirm, and perhaps even provide forums for, there is unlikely to be much change in the culture of the academy. Until graduate students become imbued with a respectfulness for the different kinds of scholarship, there is unlikely to be much change in the culture of the academy. The American Association of Agricultural Economics has an interesting record in this regard. In the mid-1980s the association leadership became concerned that the work of agricultural economists was undervalued and that the profession was not sufficiently engaging leadership and policy makers in both the public and private sector related to food, farm, and resource issues. The deficiencies were seen as affecting public and private funding for agricultural economics research, the employment of agricultural economics graduates, and the attention paid by private and public sector policy makers to the writings, opinions, and policy recommendations of agricultural economists.

In an effort to alleviate the perceived problems of public perception, in 1986 the association started a magazine entitled *Choices—The Magazine of Food, Farm and Resource Issues*. The focus of *CHOICES* was, and continues to be, twofold: 1) the demonstration that members of the profession have useful things to say about contemporary policy issues related to food, farm, and resources; and 2) the engagement of public and private sector decision-makers in dialogue about policy issues facing the food, farm, and resource sectors of the economy. In addition to the membership of the Association, the magazine is widely circulated among national and state legislatures, food and farm industry interest groups, and internationally to agricultural attachés in U.S. Embassies, among others. Its current circulation of about 5,500 reaches more nonmembers of the Association than members.

While it is difficult to know whether or not the magazine has fulfilled its dual objectives as stated above, it has had another unintended accomplishment. In the eyes of many members of the profession, the publication of an article in *CHOICES* is clearly on a par with, if not more prestigious than, the publication of an article in the Association's research journal, the *American Journal of Agricultural Economics (AJAE)*. This is true for some, if not all of the following reasons:

(1) Most in the profession read *CHOICES* while the readership of the *AJAE* is very low.

(2) Most of what is in *CHOICES* is written for the non-Association readership and is explicitly written for ease of comprehension. Formulas are not used.
(3) Articles in *CHOICES* periodically elicit letters to the editor from senior public and private sector leaders, which remind the profession that sometimes some influential decision-makers are paying attention.

Articles for *Choices* are peer reviewed but in a slightly different way than in most scholarly journals. If the editor or anonymous reviewers believe that there is an idea in the submitted manuscript to warrant further investment, the editorial staff will work with the author to craft the article to be sure the clarity and readability requirements of the magazine are achieved. None of the methodological work carried out by the profession and that dominates the *AJAE* finds its way into *Choices*. The standard for acceptance in *Choices* is the logic of the argument or evidence on the issue, the clarity of presentation, and the relevance of the issue to the society—not the profession.

What *Choices* has accomplished for the agricultural economics profession is a national standard of public scholarly writing in agricultural economics that is very much akin to extension writing. Further, that writing is given the prestige of the scholarly association, placing it on a par if not superior to publication in the association's main journal.

It is very likely that changes in the culture of the academy will not take place until there are many more publication outlets for the scholarship of integration and application that create the opportunity for extension-style writing to be seen as valid and scholarly. "Scholarly" does not mean "important" as in national in scope or significance. Scholarly means, according to Oregon State University, "original intellectual work which is communicated and the significance is validated by peers" (OSU 1999). Part of the prestige of a *Choices* article that overcomes the normal culture of the scholarly society of agricultural economists comes from the presumed validation of the article by nonacademic "peers"—the politicians and senior private and public sector managers who are more than half of the *Choices* readership. Using the Oregon State University definition of scholarship, if a university faculty member writes an extension article on a topic limited to concerns of the state or a substate region, the significance of the work must be validated by peers knowledgeable about the issue and the region—they may or may not be academics.

Imagining an engaged land-grant university must include imagining the kind of "validation by peers" embodied in the definition of scholarship employed at Oregon State University—the "peers" may or may not be academics—that's what engagement means.

An Image for the 21st Century: Engaged University and Engaged University Extension

In an earlier era, the first decade of the 20th century, President Charles R. Van Nise of the University of Wisconsin articulated what has become known as "the Wisconsin Idea"—the belief that the boundaries of the university campus are the boundaries of the state, and beyond. Van Nise, who was president of the University of Wisconsin from 1903 to 1918, declared that he would "never be content until the beneficent influence of the university reaches every family in the state" (Ward 1998, 15). According to Dr. Gerald Campbell, professor of agricultural economics and former vice chancellor of University of Wisconsin Extension, the Wisconsin Idea came out of the progressive era in Wisconsin politics.

> . . . other elements of the progressive era in Wisconsin which are associated with the "Wisconsin Idea" are things like the Legislative Reference Bureau, the Legislative Audit Bureau, and the Legislative Reference Library. These non-partisan institutions were part of the attempt to assist in creating legislation based on the best knowledge available. They were (and still are) pretty unique as elements in "informed" democracy. They remain to this day mostly non-partisan and their analysis is usually very well done. They are employers of many UW graduates and graduates of other fine institutions as well.
>
> The other dimension of this Wisconsin Idea in politics was the use of state commissions and study committees (often led by UW Professors like John R. Commons) to study an issue and recommend legislative or administrative action based on the analysis (Campbell 1999).

Sounds like President Van Nise had it right. Part of the image of *Engaged University* and of an extension system that participates in that engagement is of a university that takes the view that the borders of the state are its campus. The University of Massachusetts, the land-grant university in the commonwealth of Massachusetts, used to have a slogan that sought to embody some of that idea. It was "UMASS—Working for the Commonwealth."

The campus of Engaged University encompasses the state. As in Wisconsin under President Van Nise and in the original intent of Justin Morrill, the notion of Engaged University as a land-grant institution, includes the university participating in making the state's democracy better by having a better informed citizenry. If these are true, there are a great many things that Engaged University must be and that Engaged University Extension must do.

Imagine the following:

- Engaged University has a football team led by a coach with an advanced degree in English literature, who has personally endowed the university

library. A grade point average higher than the National Collegiate Athletic Association minimum is required to play on the team. The coach frequently goes to high schools and works with players to emphasize academics. Proceeds from the financial success of the football program accrue to the Alumni Office and its fund-raising for the total university. Shortfalls in football expenses are made up by alumni contributions.
- Leadership of the Engaged University Board of Regents comes from a prominent leader in agriculture who passionately believes that the breadth of involvement of the university in the problems of all the people of the state is a basic criterion for measuring the success of the university. University faculty positions are allocated to academic units in part as a result of performance on this measure.
- Faculty of Engaged University, whether in field assignments or on the main campus, are all associated with an academic unit that has at least some faculty members primarily engaged in discovery scholarship (commonly known as research).
- All faculty have position descriptions and report their work under a definition of scholarship that says, "Scholarship is original intellectual work that is communicated, and the significance is validated by peers."
- All faculty in Engaged University have 12-month appointments since there is an expectation that they will all contribute beyond their classroom instruction duties.
- All academic departments have faculty members who have at least a partial assignment in off-campus engagement (commonly known as extension).
- Engaged University Extension has a publication series entitled *The State of the State* to which any faculty member in any department can submit manuscripts with a suggested statewide audience. Each *State of the State* publication is a stand-alone piece directed to the audience appropriate to the content of the particular report. In the annual performance review forms used by Engaged University, *State of the State* publications are the first listed type of publication, followed by books, refereed journal articles, etc.
- Leadership of Engaged University Extension is a vice-provost with university-wide responsibility for the oversight of university engagement.
- The vice-provost for extension, in addition to oversight of the ongoing engagement activity, has responsibility for intellectual rights, copyrights, and all post-research use and control of research results. Each faculty member who completes a sponsored research project is invited to submit a manuscript for publication in the *State of the State* series. Where the source of grant funding is from a state agency, the faculty member is not just invited but expected to submit a manuscript and must explain why such a manuscript is not forthcoming, if that is the case.

- The vice-provost for research and resources is responsible for the administration of all external contract work of university personnel such as research and outreach/engagement contracts and arrangements. Obviously the individuals serving as vice-provost of research and resources and vice-provost for extension must work closely together.
- Each college or school in Engaged University has an assistant dean for extension and outreach who is responsible for that mission of the college and is responsible for the program associated with any centralized funding for that purpose that flows to the college. The assistant dean works directly with departments and faculty, particularly those with extension appointments.
- Faculty appointments in Engaged University are either 100 percent extension for field faculty, or some split between either research and teaching or research and extension for campus faculty. Faculty without extension appointments can have classroom teaching time reduced to permit outreach/engagement activity, either when internal resources and needs, or external resources and associated obligations, require it. Similar adjustments for research are not permitted since the conduct of research is more compatible with classroom instruction.
- Consulting work by university faculty is encouraged by Engaged University. The university even encourages faculty to run their consulting finances through the university office of grants and contracts, if they so choose. Where consulting proceeds go directly to enhance faculty income, none of the results of consulting work may be counted in the annual university evaluation or promotion and tenure decisions. Where the financial or other proceeds of consulting are handled by the university, and/or are used by the faculty member for professional activity including support of graduate students, office supplies and equipment, or any other professional support expense, but not for personal income, the faculty member may report such work under the "Extension/Engagement" section of the annual reporting form. Faculty with explicit extension appointments are prohibited from doing consulting within the state that results in personal income.
- Engaged University has three budget accounts from the state legislature consistent with its three missions—classroom instruction, funded research, and extension. In addition, on an ad hoc basis, state agencies make grant and contract arrangements with the university for either research or extension education in support of the particular agency's programs.
- Each office of Engaged University Extension in the counties has a set of application materials for undergraduate application to Engaged University as well as applications for, and descriptive materials about, distance and continuing education programs offered by the university.

- Each county office of Engaged University Extension has a County Extension Council. Membership on the council is an elected position in county government. Members on the council serve for three-year terms. Terms are staggered to foster "institutional memory." No council member may serve more than two consecutive terms.
- Engaged University and the state's community college systems are merged into one institution with all community college presidents reporting to the vice-provost for extension. County extension offices, when appropriate, are housed in the community college.
- The president of Engaged University, Dr. W.O. Water who has his Ph.D. in religious studies and philosophy, spends a considerable amount of his time working with public groups with interests in the programs of the university, not the least of which is the State Association of County University Extension Councils.
- Extension/outreach/service at Engaged University takes many forms. Some of the programs are carried out through university extension offices in the counties of the state. Many programs are carried out through a variety of collaborations between university faculty and other organizations. Included among the university's partners are public agencies, private-nonprofit and for-profit agencies. Many programs are delivered directly from campus faculty to the final users and collaborators on the work, while other programs have their genesis at the county level. Some county faculty members collaborate with partners across the entire state based on their local experience.
- Engaged University faculty who have responsibilities in subject matter specialties where state-level concerns involve multistate regions such as water quality, pollution, history, or other issues, are encouraged and empowered to work collaboratively with colleagues in adjacent states on both research and extension issues.
- The citizens of the state of Engagement view the Engaged University as theirs and see the county offices of University Extension as the local representative of the university. A common request to local offices is, "Do you know if anyone at the university can tell me about . . . ?" The frequent answer is, "I'm not sure, but we'll sure help find out. Let me get the university directory online and we'll explore."
- State government agencies and state legislators understand that as an academic institution Engaged University is quite different from other state-financed and state-responsible organizations, and not nearly as hierarchical in its organization. Leadership of the university makes no pretense of speaking on behalf of faculty for work on behalf of state government. However, they do work hard at helping to broker collaborations between talented faculty and government needs for assistance. Thus the image that university leaders convey of the

university to state agencies and the legislature is of an open pit mine where there is much to be gained by some careful digging around to locate the right ore. The office of the provost or either of the vice provosts can direct requests to the right college or even the right department so that legislators or agency personnel can more quickly locate the talent they want. A key word subject index is available to the university telephone operator and also on the university Web site to direct inquiries to specific faculty members.

- On an annual basis, Engaged University publishes a report entitled *What We Heard and What We Did: Annual Report of Engaged University to the People of the State.* The report is explicitly about the character of the previous year's activity beyond the formal degree instructional programs of the university. The first part of the report is a statement about the character of the problems that people of the university considered important to the people of the state at the beginning of the period. The remainder of the report details the responses to the perceived problems as well as a synopsis of ongoing outreach programs. The report lists the title, author, and distribution including numbers of *State of the State* reports issued during the year. It reports the numbers of volunteers involved in the various extension programs. It lists all of the departments of the university actively involved in the engagement. It lists the myriad of organizations and agencies with which the university partnered in its engagement. Clearly the validity of the report rests on both the congruence between the perceived problems and pressing societal issues, and the efficacy of the university responses.
- Engaged University has a policy that any faculty member invited by the state legislature to give leadership to a major nonpartisan study task force or commission of the legislature may be supported half time in that activity with university resources for one year. If the work requires full-time commitment, the legislature must pay the other half salary and benefits during the first year, and all compensation beyond the first year. The legislature is aware of that university policy.
- Engaged University seeks to have its communications staff organize and prepare a major policy education piece in the form of a newspaper tabloid on some major policy question facing the state each year. The tabloids are patterned after the Oregon State University tabloids on salmon and poverty. They are seen as a way to educate university faculty and the citizens of the state to the particular issue, and to a process of public deliberation that explicitly recognizes that there are many points of view on many important public policy issues that are deserving of respect.
- Engaged University Extension has an Advisory Council, which is comprised of representation from the State Association of County

University Extension Councils, and representatives of partners of the university in its engagement. Half the members of the Advisory Council is from the Association of County University Extension Councils and the other half is from partners of the university in other dimensions of its engagement. There is at least one person on the Advisory Council nominated by each of the university's colleges or schools. The University Extension Advisory Council has at least one of its members on the University board of regents.
- Engaged University has a strategic plan that anticipates the kinds of faculty needed and the job descriptions. The plan is worked out and legitimized by faculty and departments such that continuity of direction can be achieved even as administrators change throughout the university.

Conclusion

The Kellogg Commission report, *Returning to Our Roots–The Engaged Institution,* has a seven-part test of guiding characteristics for an engaged institution. They are:

(1) Responsiveness
(2) Respect for partners
(3) Academic neutrality
(4) Accessibility
(5) Integration
(6) Coordination
(7) Resource partnerships (Kellogg Commission 1999, 29)

Engaged University, as described above, seeks to meet all of these tests and Engaged University Extension is the instrument for facilitating many of the requirements.

What are missing in the vision above are the issues over which a single university has little control. This point is important in that some of the control of extension and the land-grant universities into the 21st century are well beyond the control of any single university and its leadership. Of particular note is the importance of the scholarly societies. Fundamentally they are more important than the annual scramble over roughly a half billion dollars for extension and a half billion dollars for agricultural research that explains most of the USDA/land-grant and NASULGC constellation of forces and activity.

In the absence of control over external forces such as scholarly societies, university leadership must be very clever in suborning the negative influence of external forces, while affirming the positive influences. Much of that must be accomplished by the manipulation of symbols rather than by the issuance of edicts or directives. For example, the

Engaged University makes a significant statement about engagement when it places the *State of the State* publications—a category of publication that is outreach and engagement education in its intention, accessible in its writing style, and inviting of critique from nonacademic peers—at the head of the list of evidence of scholarly activity accredited by the university. There will be many well-regarded university scholars without that premiere publication in their annual reports. If the honor of being named a University Distinguished Professor also carried the requirement of having a *State of the State* publication, the force of the symbolism would be even more complete.

In its representations to national organizations like NASULGC and to the federal legislators from its state, Engaged University takes the position that part of the original social contract between the people and the land-grant universities was to bring the university talents to bear on the problems of the people. In the early years of the land-grant colleges, the funding for solving the problems of the people was initially from federal Hatch funds with extension funds from Smith-Lever. At the turn of the 21st century, the research portfolio of the land-grant universities is huge. In 1996, the top 43 land-grant universities received $3.7 billion dollars in support of federal research and development alone (Chronicle 1998, 34). Except for the relatively small amount of money (roughly a half billion dollars) for extension from the USDA, there is little or no matching of extension support associated with any of the other sources of federal funding. Accordingly, the position of Engaged University is: If the people of the nation have a claim on the results of research funded by the federal government, that claim should be empowered and funded in terms of outreach support associated with the research. Each research grant awarded from a federal agency should have an associated funded extension obligation to extend that knowledge to the citizens affected or best able to make use of the research results.

There is no claim that the imagining in the previous pages of this chapter is realistic. One is reminded of the John Lennon lyrics that assert that imagining is easy, if you try. Unfortunately, bringing the imagined images of an engaged land-grant university—whether the one in the pages above or in the Kellogg Commission report—into being is not at all easy, but will surely be worth the effort. Perhaps one day there will be some land-grant universities that are once again "people's universities," and some of America's preeminent institutions of higher education will have entered, once again, into a social contract with America. The chorus to the Lennon song says, "You may say I'm a dreamer, but I'm not the only one," which is fortunate indeed.

10

Conclusion—Renegotiating or Abandoning a Social Contract

A common reaction to "I'm writing a book on the future of extension and land-grant universities" was "Do they have a future?" A good friend, perhaps on a bit of a blue day, said that perhaps the book would be an epitaph for the land-grant universities as instruments of social change in American society. If the book is to be an epitaph because the land-grant universities cannot again be instruments of social change, perhaps the book will at least chronicle some of the reasons for their demise in that role. The difference between an epitaph and a plan for action is principally in the response of the system to the details of the circumstance described.

If the land-grant universities generally, or any one of them specifically, fails to make the necessary changes to again become people's universities, and if extension at those institutions fails to play a significant role in that change, it will not—repeat, will not—be because of the leadership of the extension programs. There are too many carcasses of dedicated extension directors and other leaders who tried to change the programs and got shot down or dismissed in the process, to in anyway hold those folks responsible for the demise. Indeed, extension directors will be among the saddest at the funeral.

Deans of agriculture bear a great responsibility in the extension wars for their failure to act as statesmen in moving extension into an administrative relationship where university presidents, chancellors, and provosts are forced to more directly assume responsibility for the character of its program. Deans of agriculture also bear considerable responsibility for not creating forthright and mutually respectful relationships with their farm commodity groups, bringing them along to the understanding of why changes are necessary in the way they do business generally and about the future of extension, specifically. Universities want and need to be engaged, but being taken hostage is another thing.

Even where extension is not under their control, deans of agriculture should take on a partial extension appointment where the audience is the leadership of farm and agribusiness organizations. The program they organize should be directed at helping agricultural leaders understand what is in their own, long-run best interest. Such a program will certainly include helping agricultural leaders understand that as incomes in the society rise, there is not a commensurate rise in demand for food, but there is a greater rise in the demand for environmental quality. The comparison means that in conflicts between environmentalists and farmers, farmers will almost always lose the battle, and farmers will do better to make friends with the environmental interests than fight them. The deans' extension program may also include suggesting that it will be in farmers' best interest, too, if extension in the county provides programs for a wide array of audiences such that there is a broad base of support in the community for the total program, including the agricultural program. In terms of the hypothetical history extension program discussed in earlier chapters, the farm leadership in counties should form a coalition with the historical society in support of extension, assuming there is a history extension program. If there is not a history extension program, the farm groups should explore whether there should be one.

Furthermore, there is more to successful engagement programming than simply getting leadership of the engagement out from under the deans of agriculture. Indeed, in some places, extension remains under the deans of agriculture and the leadership of engagement is elsewhere. Sometimes that is the best that can be accomplished, given the forces at work in the system. It may be a useful interim step and is perhaps what is taking place at Pennsylvania State University, where the director of extension is both associate dean of agriculture and associate vice president for outreach. However, where such arrangements are seen as final, it does not augur well for either the total engagement effort or for the long-term vitality of extension.

Chancellors, presidents, provosts, and other senior university leaders, including regents, trustees, curators, visitors, or whatever such oversight groups are called, must make engagement an equal and integrated part of the university program. The Kellogg Commission Report *Returning to Our Roots: The Engaged Institution* (1998) is a first-class document for articulating its importance. It does not, however, tell much about how to do it or about the circumstances that mitigate against engagement happening. University leaders—all the folks mentioned above—must attend to creating the culture of the campus and of the university within the state such that they enable engagement. Some of that work will be in establishing administrative lines of authority, and part will be in attending to the politics of extension and engagement in both the research and outreach domains. Much of the work of such people is far from the domain of management and is more in the domain of manipulating symbols and creating metaphors, though creating and changing some of the symbols or metaphors will require tough management decisions.

For example, in the imagined Engaged University of the previous chapter, all faculty members have 12-month appointments because the symbolism of 12-month appointments is much more appropriate to an engaged university than are 10-month appointments. Given the history of most land-grant universities, a switch to year-long appointments for the faculty who now carry 10-month appointments would take huge amounts of leadership capital and might indeed be the kind of issue over which a university leader loses a job. However, in the absence of making that change, the discrepancy between those with year-long appointments and those without will have some negative impact on creating a culture where engagement is the responsibility of all faculty. In truth, the current academic year appointments are essentially license for those faculty members to earn additional income during the remaining "unpaid" months and are considered an inducement to get grants and contracts that will yield the additional income. It is true that some of that contract/grant activity may be "engagement."

Alternatively, consider the manipulation of the symbolism within Engaged University of what counts most in the evidence of scholarly achievement. By creating a publication series directed to the people of the state, and then anointing it by making it the highest priority in reporting scholarly achievement, the university leadership does get to participate in the discussion of what is good scholarship and what is not good scholarship. At individual institutions, that may even invade the culture of graduate education and what graduate students consider to be scholarly excellence. Such decisions are usually left to the scholarly societies and their departmental representatives, subject matter by subject matter. It seems to this writer that university presidents and other university leaders should bring the agenda of engagement directly to the scholarly societies at least by moving the subject on to the agenda of the National Academy of Science.

Changing university culture is very difficult to accomplish and can be very slow indeed. In conversation with Dr. James Ryan, vice president for Outreach and Cooperative Extension, Pennsylvania State University, Ryan (1999) quipped that it was akin to realigning the stars. According to President Emeritus John Byrne, the person most responsible for the changes at Oregon State University, "style and process are everything" (Byrne 1999). Unless everyone who wants to have a say on an important matter gets a chance to speak, there will not likely be much sustainable change. But giving interested parties a chance to speak to an issue is different than giving them control or allowing them to take you hostage. In the case of land-grant universities, the folks who must be taken into consideration on the changes in the university are not only the faculty, but also the leadership of partner organizations and prominent interest groups. University leaders who are uninterested in having such folks attended to in the name of the university will not likely be able to change the culture.

But a supportive university culture only creates the setting in which engagement can widely and easily occur—it does not cause it to happen. That must come from the creative initiative of many county and campus faculty and staff who see opportunities, are sufficiently attentive to hear people's needs, or systematically collect evidence of needs and are creative in their response. Engagement can take many forms. However, the truly effective engagement, such as some of the examples in Chapter 8 on Promises and Possibilities, involve considerable directed effort and investment. Engagement, regardless of its form, is time-consuming and demanding. If it is to be truly engagement, as in challenging the supporting disciplines as illustrated in Chapter 4, then there is a further obligation for those engaged to write about those challenges for their discipline.

If any particular kind of university engagement is to result in solving someone's problems or in producing a more informed citizenry and a greater appreciation of the role of the university, there must be some evidence of the institution's involvement beyond the contribution of the scholar/staff. This is the stuff of the "attribution condition" necessary for gaining support from people served as discussed at length in several earlier chapters. It means that the university seeking to be engaged must commit considerable resources to publications in support of the effort.

By any stretch of the imagination, a tabloid on salmon policy in Oregon as an insert in local newspapers distributed to 650,000 households in a state population of 3.3 million, is a better public relations statement about the university than the slickest PR promo piece Oregon State University could possibly imagine. This is true primarily because the tabloid is itself good education. The OSU tabloid, *A Portrait of Poverty in Oregon*, has had 850,000 copies first run. Computer/information technology people who argue that paper publications are obsolete and electronic publication is all that is necessary do not understand the role of paper publications beyond providing information and education. Paper publications are important in garnering credit and establishing turf.

As mentioned in the previous chapter, engagement, which in this book means renegotiating a social contract with the people who sponsor the university, should be measured not only by the character of the involvement with the people, but also by the breadth of that involvement across the university. What that also means is that the involvement of different parts of the university will take different forms. One should not get too hung up on a particular form or style of engagement, since attributes of problems and solutions to people's problems will dictate the kind of involvement. Attributes of the things people relate to have a profound effect on the character of the relationship between the people, and presumably also between the people working for a university and the people in the society. Thus, it is that the character of the engagement in the physical or engineering sciences will be quite different from the character of the engagement in the social science or the humanities. It will be quite different if the group the university is engaging is looking for assistance in the business part of their lives or in the family part of their lives. It will be different if the businesses seeking assistance are large corporations or if they are small, family-run businesses like hair salons and farms.

No matter how much a particular model of engagement is presumed to be most effective, the method chosen and finally used will be determined by the attributes of the information, the audience, and the problem being dealt with. Where the issue requires a primarily technological solution, the engagement will look much like technology transfer. Where the issue is assisting a community deal with a problem and the first steps are to determine the extent of the problem in the community, community-based collaborative research and its transformational learning advantages will likely obtain.

Where research or development scholarship result in products or processes that can be patented, they will be—either by the private firm that funded the work, the university, or the scholar. Many of the products of technological advance out of the physical and biological sciences have private good attributes, and that pretty much dictates how they will be used. Where such product development is publicly funded or funded from other university resources, failure to restrict the subsequent use of the knowledge via patents may only lead them to be stolen and exploited via patents by organizations that never paid for their development. Licensure of such products by the university may be the most effective way to share the knowledge with the society, and also recover some of the proceeds of the commercialization of the product for the university. For some, this license or other release through the private sector will qualify as engagement.

Such arrangements are usually thought to be well beyond the scope of extension types of programming; however, there is a long history of extension sponsoring or spinning off businesses that are built around information products that have private good attributes. The development of Dairy Herd Improvement Associations as cooperative businesses around improved dairy cattle semen and artificial insemination is but one example. The example of textbook development in the Humanities Extension Program in North Carolina described in Chapter 8 has some of the same characteristics. Those who would deal extension leadership out of the disposition of results of research and intellectual property rights make an error. The North Carolina State University textbook example suggests that perhaps even greater involvement of the university in the utilization of its patented products and intellectual properties may make some sense to facilitating the continued engagement.

Engagement means staying attuned to the issues faced by people. If that means retaining to the university a resource of its creation that facilitates staying attuned to the society, then there is no particular reason to "spin it off" unless the job is done and there is no further input the university can make or insight the university can garner.

Increasingly, engagement in the physical and biological sciences is by direct or indirect relationships with corporate interests who fund research and then expect to be able to exploit the results commercially. It is thus likely that the knowledge base that will increasingly be called upon for the type of engagement that leads to extension/outreach programs with groups of people or communities will be in the social sciences, human or behavioral sciences, or the humanities. Clearly that is the character of the portfolio of the University of Wisconsin Extension, with its strong emphasis on rural/community/economic development.

The evidence of greater need for economic and other social science input into agricultural extension is also apparent as farmers deal with state and local policy and seek to position themselves strategically for the future. However, the failures in the political marketplace are such that farm groups will not support that change—they are still preoccupied with on-the-farm production technology and its management. That is one price farmers pay for taking extension hostage, as discussed in Chapter 6. Farmers will get what they want but not what they need. A separate renegotiated social contract with agricultural interests, via agricultural deans' working directly with them as discussed above in this chapter, will be required to change that.

There already is a large amount of engagement between the society and land-grant universities. Nothing said in this book should be interpreted as indicating otherwise. Unfortunately, much of the engagement occurs in spite of, rather than because of, policies, practices, and culture in the universities, including the engagement facilitated and encouraged by extension. Extension, as the most logical instrument for facilitating engagement in land-grant universities, is in many places unable to serve the entire university.

In many places where engagement has occurred, whether inside or outside of the aegis of extension, it happens because of almost heroic behavior by campus faculty members who accept a professional handicap as part of the price of doing the work. Similarly, within extension, successful programming at the county level occurs in spite of, rather than because of, the university, and often because of heroic behavior by county staff. For campus faculty members who work within the aegis of extension, but are dealing with subject matter or a program area that is not in the mainstream of the existing extension program, the problems and disincentives are only slightly less discouraging than for faculty who do not come under extension.

Renegotiating a social contract with the American people by the land-grant universities means solving the problems of extension and the universities such that the normal instincts of many faculty members within the university to have their scholarship make a difference in people's lives, can easily be achieved. The Kellogg Commission on the Future of State and Land-Grant Universities report, *Returning to Our Roots—The Engaged Institution,* is a bold statement about why it is important to renegotiate the social contract, but it does not say much about what stands in its way and what must be overcome. We have tried to detail some of that in this book. The task is difficult; the influences are many, diffuse, and elusive—it's like trying to realign the stars.

The politics of the land-grant university is county politics, state politics, federal politics, and scholarly society politics in addition to campus politics. Where extension has successfully extracted itself from the suffocating grasp of agricultural interests, it appears to have been solved at the level of county politics and collective county support at the state level. The dilemma imposed on the system by the U.S. Department of Agriculture as the federal partner is described in some detail in Chapter 7, yet the prospects for significant change at that level are hardly promising. Similarly, there are few examples where scholarly societies have in any way come to grips with renegotiating the evaluation of scholarship. The policy levers available to university leaders are few and small, as evidenced at Oregon State University. When described in this way, one is led to wonder how any engagement occurs at all, making the persistence of those who do it well that much more impressive.

Plan of Action or Epitaph? Renegotiating or abandoning a social contract? The answer is not obvious. Perhaps this analysis cum criticism may help to weigh the scales on the side of renegotiating a new social contract for the 21st century by these institutions. During much of the 20th century, the land-grant universities had achieved the apex of what higher education could be in the world. However, in the latter half of the century, they have seemed to have lost their way; a part following and aspiring to the images of the private institutions, and the other part—the core land-grant colleges—stuck in their agricultural past and holding extension there with them.

It has been their engagement as people's universities that made the land-grant universities better than Harvard, Yale, Stanford, Humbolt, Cambridge, or Oxford in renewing culture, interpreting the past, and expanding our understanding of the human condition. Unless they carry out that renegotiation and return to their roots, they stand in danger of being no better. From their beginnings, in the values of American democracy, the land-grant institutions were to be better than the elite institutions and were to make the democracy itself better, in part on the basis of whom they admitted to their classrooms. Now they must achieve their greatness on the basis of how much of the university is engaged with America and with whom they engage. There is much to be done in renewing and fulfilling the land-grant universities' social contract with America into the 21st century.

References

Agricultural and Applied Economics. 1999. Statement in public meeting to CSREES Review Team for the Department of Agricultural and Applied Economics, 6 June, at Virginia Polytechnic Institute and State University, Blacksburg, Virginia.

Ahearn, Mary. 1999. Unpublished data from CSREES/USDA by way of the Economic Research Service/USDA.

Alston, Julian M. and Philip G. Pardey. 1996. *Making Science Pay: The Economics of Agricultural R&D Policy.* Washington, D.C.: The AEI Press.

Barkley, Paul W. 1984. Rethinking the Mainstream. *American Journal of Agricultural Economics* 66(5).

Batie, Sandra. 1991. Personal conversations.

Beneke, Raymond R. 1983. August 23 Convocation Address. *Department of Economics Newsletter,* December. Iowa State University.

Benning, Linda. 2000. Personal written communication, 6 February.

Benson, Esra Taft. 1954. Secretary of Agriculture Memorandum No. 1368, Activities of Department Employees With Relation to General and Specialized Organizations of Farmers. Washington, D.C.: U.S. Department of Agriculture, 24 November.

Blank, Steven C. 1998. *The End of Agriculture in the American Portfolio.* Westport, Connecticut, London: Quorum books.

Blank, Steven C. 1998a. Personal written communication, 19 August.

Blaug, Mark. 1980. *The Methodology of Economics.* Cambridge: Cambridge University Press.

Bloome, Peter. 1999. Personal written communication, 28 October.

Blyth, Dale. 1999. Personal conversation at University of Minnesota, 23 September.

Board on Human Sciences. 1999. Project Inventory, Fiscal Year 1998. Washington, D.C.: NASULGC.

Bok, Derek. 1990. *Universities and the Future of America.* Durham and London: Duke University Press.

Bonaparte-Krogh, Paul M. 1999. Executive Director, Tompkins County Extension, Ithaca, New York. Personal conversation, July.

Bonnen, James T. 1986. A Century of Science in Agriculture: Lessons for Science Policy. Fellows Lecture, *American Journal of Agricultural Economics* 68(5).

Bonnen, James T. 1999. Personal written communication, February.

Boulding, Kenneth E. 1975. Quality Versus Equality: The Dilemma of the University. *DEADALUS* Winter 1975. Boston: The American Academy of Arts and Sciences.

Bowen, Howard R. 1996. *Investment in Learning—The Individual and Social Value of American Higher Education.* San Francisco: JosseyBass Publishers.

Boyer, Ernest L. 1990. *Scholarship Reconsidered—Priorities of the Professoriate.* Princeton: The Carnegie Foundation for the Advancement of Teaching.
Boyer, Ernest L. 1997. Forward, *A Classification of Institutions of Higher Education,* The Carnegie Foundation for the Advancement of Teaching.
Bromley, Daniel and Richard McGuire. 1991 Debate: Bromley, Daniel. Technology, Technical Change, and Public Policy: The Need for Collective Decisions. McGuire, Richard. Food, Energy, and Environmental Quality: The Necessity For Balance. *CHOICES,* Second Quarter.
Burns, Leslie Davis. 1999. The Impact of the 1995 Promotion and Tenure Guidelines and Revised Definition of "Scholarship" on the OSU Campus. Corvallis, Oregon: Oregon State University, July.
Busch, Lawrence and William B. Lacy. 1983. *Science, Agriculture, and the Politics of Research.* Boulder, Colorado: Westview Press.
Byrne, John. 1999. Personal conversation, 16 December.
Campbell, Gerald. 1999. Personal written communication, 30 November.
Carnegie Foundation for the Advancement of Teaching. 1967. The University at the Service of Society. 1967 Annual Report, New York.
Carter, Charles. 1980. *Higher Education for the Future.* Oxford, England: Basil Blackwell Publisher.
Castle, Emery N. 1981. *Agricultural Education and Research—Academic Crown Jewels or Country Cousin?* Washington, D.C.: Resources for the Future, Inc., March.
Castle, Emery N. 1989. Is Farming a Constant Cost Industry? *American Journal of Agricultural Economics* 71(3).
Castle, Emery N. 1993. On the University's Third Mission: Extended Education. Corvallis, Oregon: Oregon State University.
Chronicle of Higher Education. 1998. Almanac Issue 1998–1999 Vol. XLV(1).
Chubin, Daryl E. and Edward J. Hackett. 1990. *Peerless Science.* Albany: State University of New York Press.
Clark, James W. Jr. 1999. Extending the Humanities and Social Sciences in North Carolina. Raleigh, North Carolina. Unpublished paper, Humanities Extension/Publications, North Carolina State University, August.
Clark, James W. Jr. 1999a. Personal communication, 9 September.
Clinton, Hillary Rodham. 1996. *It Takes a Village: And Other Lessons Children Teach Us.* New York: Simon & Schuster.
Cole, J., E. Barber, and S. Graubard, eds. 1994. Balancing Acts: Dilemmas of Choice Facing Research Universities, in *The Research University in a Time of Discontent.* Baltimore and London: The Johns Hopkins University Press.
Commonwealth of Virginia. 1995. *Acts of the General Assembly (Regular Session).*
Cooperative State Research, Education, and Extension Service (CSREES). 1999. Online. Internet. 13 July 1999. Available: http://www.reeusda.gov/1700/about/csreesa2.htm#mission.
Crichton, Michael. 1990. *Jurassic Park.* New York: Knopf, Distributed by Random House.
Cross, Lawrence H. 2000. Myths and Facts Regarding the SOL Reform Movement. Unpublished discussion paper. Blacksburg, Virginia, Department of

Educational Leadership and Policy Studies, Virginia Polytechnic Institute and State University, January.
Cross, Lawrence H. 2000a. Personal conversation, 6 March.
Dixon, Marlene. 1976. *Things Which Are Done in Secret.* Montreal: Black Rose Press.
Dooley, John. 1998. Defining the Mission of Virginia Cooperative Extension: A Policy Analysis Based Upon a Historical Study. Blacksburg, Virginia. Ph.D. Dissertation in Educational Administration, Virginia Polytechnic Institute and State University.
Drabenstott, Mark. 1999. New Futures for Rural America: The Role for Land-Grant Universities. Hatch Memorial Lecture, NASULGC Meetings, San Francisco, 8 November.
Evenson, Robert E. and Yoav Kislev. 1975. *Agricultural Research and Productivity.* New Haven: The Yale University Press.
Ewert, Merrill. 1999. Personal written communication, 16 September.
Feld, B.T. 1975. On Legitimizing Public-Service Science in the University, *DAEDALUS,* Winter 1975. Boston: The American Academy of Arts and Science.
Feldman, Kenneth A. 1987. Research productivity and scholarly accomplishment of college teachers as related to their instructional effectiveness: A review and exploration. *Research in Higher Education,* 26(3):227. *Journal of the Association for Institutional Research.* New York: Agathon Press, Inc.
Feller, I.L. Kaltreider, P. Madden, D. Moore, and L. Sims. 1984. Overall Study Report: Findings and Recommendations Vol. 5, *The Agricultural Technology Delivery System: A Study of Agricultural and Food Related Technologies.* Prepared for Science and Education, U.S. Department of Agriculture. University Park, Pennsylvania, Institute for Policy Research and Evaluation, The Pennsylvania State University.
Feyerabend, Paul. *Science in a Free Society.* 1978. London, NLB.
Foil, Rodney. 1999. Retired Vice President for Agriculture, Mississippi State University. Personal communication, July.
Garvin, David A. 1980. *The Economics of University Behavior.* New York: Academic Press.
Gasset, José Ortega y. 1944. *Mission of the University.* Princeton, N.J.: Princeton University Press.
Glassick, Charles E., Mary Taylor Huber, and Gene I Maeroff. 1997. *Scholarship Assessed—Evaluation of the Professorate.* San Francisco: Jossey-Bass Publishers.
Glion Declaration: The University at the Millennium. Online. Internet. November 1999. Available: http://www.weberfamily.ch/glion/.
Graubard, Stephen R. 1997. *DAEDALUS, The American Academic Profession,* Fall, p. v, vi. Boston: The American Academy of Arts and Sciences.
Havelock, R.G. 1969. *Planning for Innovation Dissemination and Utilization of Knowledge.* Ann Arbor: Institute for Social Research, University of Michigan.
Hildreth, R.J. and Walter J. Armbruster. 1981. Extension Program Delivery—Past, Present and Future: An Overview, *American Journal of Agriculture Economics,* American Agricultural Economics Association, December, p. 853.

Hoard's Dairyman. 1991. Agricultural Extension Is Under Attack, Editorial, *Hoard's Dairyman*, 10 February.
Hoch, Irving. 1984. Retooling the Mainstream. *American Journal of Agricultural Economics* 66(5).
Houglum, Lyla. 1999. Personal written communication, 6 October.
Houglum, Lyla, 1999a. Personal written communication, 1 December.
Huffman, Wallace E. and Robert E. Evenson. 1993. *Science for Agriculture*. Ames: Iowa State University Press.
Jaschik, Scott. 1991. Political Activists Working to Change Land-Grant Colleges. *The Chronicle of Higher Education*. 20 March.
Johnson, Glenn L. and Lewis K. Zerby. 1973. *What Economists Do About Values*. East Lansing: Department of Agricultural Economics, Center for Rural Manpower and Public Affairs, Michigan State University.
Johnson, Stanley R. 2000. Vice Provost for University Extension, Iowa State University. Personal conversation. January.
Kammann, Nan. 1999. St. Louis Storytelling Festival. Online. University of Missouri-St. Louis Continuing Education & Outreach. Internet. 1 October 1999. Available: http://www.umsl.edu/~conted/storyfes.htm.
Katouzian, Homa. 1980. *Ideology and Method in Economics*. London: The MacMillan Press, Ltd.
Kellogg Commission. 1998. *Returning to Our Roots: The Engaged Institution*, Kellogg Commission on the Future of State and Land-Grant Universities. Washington, D.C.: National Association of State Universities and Land-Grant Colleges.
Kellogg Presidents' Commission. 1996. Joint Statement, Kellogg Presidents' Commission on the 21st Century State and Land-Grant University, E. Gordon Gee (Chairman), President, The Ohio State University; Dolores Spikes (Vice-Chairwoman), President, Southern University System; John V. Byrne, (Director), President, Oregon State University; C. Peter Magrath, President, NASULGC. January. Washington, DC: National Association of State Universities and Land-Grant Colleges. Online. Internet. 19 April 2000. Available: http://www.nasulgc.org/Kellog/STATEMENTS/comstate.html.
Kile, Orvelle Merton. 1948. *The Farm Bureau Through Three Decades*. Baltimore: Waverly Press.
Kuhn, Thomas S. 1970. Logic of Discovery or Psychology of Research? in *Criticism and the Growth of Knowledge*. Lakatos and Musgrave, ed. Cambridge: Cambridge University Press.
Lande, John. 1994. *Teen Assessment Project—Five Year Impact Report, 1989–1994*. Madison: University of Wisconsin-Extension.
Lerner, Richard M. and Lou Anna K. Simon. 1998. Editors, *University-Community Collaborations for the Twenty-First Century*. New York and London: Garland Publishing, Inc.
Lewis, Lionel S. 1975. *Scaling The Ivory Tower*, Baltimore: The Johns Hopkins University Press.
Magrath, C. Peter. 1996. Statement on the Kellogg Commission on the Future of State and Land-Grant Universities, 30 January. Washington, D.C.: National

Association of State Universities and Land-Grant Colleges. Online. Internet. 19 April 2000. Available: http://www.nasulgc.org/Kellog/ANNOUNCEMENTS/cpmstm.html.

Martin, Brian, 1981. The Scientific Straightjacket, in *The Ecologist,* January-February.

Mayer, A. and J. Mayer. 1974. Agriculture, the Island Empire. *DAEDALUS* 103. Boston: The American Academy of Arts and Sciences.

McDowell, George R. 1985. The Political Economy of Extension Program Design: Institutional Maintenance Issues in the Organization and Delivery of Extension Programs. *American Journal of Agricultural Economics,* 67(4).

McDowell, George R. 1987. Why Many Extension Economists Are Not at the Cutting Edge and What They Can Do about Moving the Edge. Paper prepared for American Agricultural Economics Association, Pre-Conference Extension Workshop, "Maintaining The Cutting Edge." East Lansing, Michigan, 31 July–1 August. Unpublished.

McDowell, George R. 1991. The USDA and The Extension System Revisited or If You Haven't Visited Extension Recently, You Better Do It Soon, Cause It Isn't Going To Be There Long. Presentation to the National Workshop for Extension Agricultural Program Leaders, Nashville, Tennessee, 3–5 April. Unpublished.

Michigan State University Extension (MSUE). 2000. AoE Area of Expertise Teams. Online. Internet. 22 February 2000. Available: http://www.msue.msu.edu/aoe/.

Miller, Paul A. 1961. The Agricultural Colleges of the United States: Paradoxical Servants of Change. Centennial Convocation, American Association of Land-Grant Colleges and State Universities.

Miller, Paul. 1999. Personal written communication, Columbia, Missouri, 28 April.

Minnesota Extension Service. 1999. *Minnesota Impacts!* Database of the Minnesota Extension Service, September.

Morrill, Justin Smith. 1887. I Would Have Higher Education More Widely Disseminated. Address delivered at the Massachusetts Agricultural College, 1887. Reprinted 1961, Amherst, Massachusetts: University of Massachusetts Centennial Committee.

Morse, George. 1999. Personal conversation, April.

National Education Association. 1984 *The NEA 1984 Almanac of Higher Education.* Washington, D.C.: National Education Association.

National Science Foundation. 1999. Directorate of Biological Sciences, Plant Genome Research Program. Online. Internet. 15 July 1999. Available: http://www.nsf.gov/bio/dbi/dbi_pgr.htm.

National Science Foundation. 1999a. *Program Solicitation, NSF 99–125,* National Science Foundation, Directorate for Mathematical and Physical Sciences, Division of Materials Research, Materials Research Science and Engineering Centers. Online. Internet. July 1999. Available: http://www.nsf.gov/cgi-bin/getpub?nsf99125.

Nipp, Terry. 1999. Personal communication, 8 July.

Nobel, Fraser. 1979. Relevance of the Curriculum to the Needs of Society in *Pressures and Priorities—Report of the Proceedings of the Twelfth Congress of the Universities of the Commonwealth*, London: The Association of Commonwealth Universities.

Nunnally, Richard. 1999. Personal written communication, 28 May.

Olsen, J.L. and C.D. Boyer. 1999. Integrating extension field faculty into academic homes: the Oregon State University experience. Presented paper, American Society for Horticultural Science Annual International Conference, Minneapolis, Minnesota, July.

Olson, Mancur, Jr. 1968. *The Logic of Collective Action*, New York: Schocken Books.

Oregon State University. 1999. Faculty Handbook. Online. Internet. 18 August 1999. Available: http://osu.orst.edu/staff/faculty/handbook/toc.htm.

Pease, James. 1999. Conversation with Dr. James Pease, Farm Management Extension Specialist, Department of Agricultural and Applied Economics, Virginia Polytechnic Institute and State University, Blacksburg, Virginia. March.

Peri, Marianne, Zhongren Jing, Roy Pearson, Joel D. Sherman, Thomas D. Snyder. 1997. *International Education Indicators: A Time Series Perspective*. Washington, D.C.: National Center for Educational Statistics, U.S. Department of Education.

Peters, Scott Joseph, 1998. Extension Work as Public Work: Reconsidering Cooperative Extension's Civic Mission. Ph.D. Dissertation, University of Minnesota, Minneapolis.

Peterson, Robert G. 1975. Higher Education's Social Contract To Serve the Public Interest, *Education Record*, Fall 1975. Washington: American Council on Education.

Press, Eyal and Jennifer Washburn. 2000. The Kept University. *The Atlantic Monthly*, 285(3). Boston: The Atlantic Monthly Group.

Prosise, Everette. 1999. Personal conversations, June.

Pulver, Glen. 1998. Personal written communication, 21 October.

Rainsford, George N. 1972. *Congress and Higher Education in the Nineteenth Century*, Knoxville, Tennessee: The University of Tennessee Press.

Rankin, Richard. 1999. Personal communications, 22 July and 8 September.

Rasmussen, Wayne D. 1989. *Taking the University to the People, Seventy-five Years of Cooperative Extension*. Ames: Iowa State University Press.

Riese, Thomas. 1999. Personal written communication, September.

Riley, David. 1999. Presentation to National Public Policy Education Conference, St. Paul, Minnesota, September.

Riley, David. 1999a. Personal written communication. 30 September.

Ryan, James H. 1999. Personal conversation. State College, Pennsylvania, 17 February.

Schaller, Neil. 1999. Personal communication, 1–3 August.

Schaller, Neil. 1999a. Personal communication, 16 August.

Schertz, Lyle. 1999. Former editor of *CHOICES*. Personal conversation, 1–3 August.

Schnittker, John. 1999. Former Assistant Secretary of Agriculture. Personal conversation, 1–3 August.
Schnittker, John, 1999a. Personal conversation, 6 August.
Schuh, Edward G. 1986. Revitalizing Land-Grant Universities—It's Time To Regain Relevance. *CHOICES*. AAEA, 2nd Quarter.
Scott, Roy V. 1970. *The Reluctant Farmer—The Rise of Agricultural Extension to 1914.* Urbana: University of Illinois Press.
Shapley, Deborah and Rustum Roy. 1985. *Lost at the Frontier*, Philadelphia: ISI Press.
Small, Stephen. 1998. Personal written communication, November.
Stephens, Michael D., and Roderick, Gordon W., editors. 1975. *Universities For A Changing World The Role of the University in the Later Twentieth Century*, Newton Abbot, England: David and Charles.
Swackhamer, Gene. 1999. Personal conversation, February.
Swanson, Earl R. 1984. The Mainstream in Agricultural Economics Research. *American Journal of Agricultural Economics* 66(5).
Swiger, L.A. 1993. Dean, College of Agriculture and Life Sciences, Virginia Polytechnic Institute and State University. Personal communication, 5 January.
Swiger, L.A. 1998. Dean, College of Agriculture and Life Sciences, Virginia Polytechnic Institute and State University. Personal conversation, September.
Taylor, John F.A. 1966. *The Masks of Society.* New York: Appleton-Century-Crofts: Meredith Publishing Company.
Taylor, John F.A. 1981. *The Public Commission of the University.* New York, New York: University Press.
Torgersen, Paul. 1994. Personal conversation, March.
Turner, Ronald J. 1980. Personal letter to Emily Thach, 19 March.
Turner, Ronald J. 1999. Personal written communication, 1 October.
Tweeten, Luther. Personal conversation, 1–3 August 1999.
Tymoczko, Maria. 1999. Will the Traditional Humanities Survive in the 21st Century. Paper presented at Re-Organizing Knowledge: Transforming Institutions, Knowing, Knowledge and the University in the XXI Century, University of Massachusetts, 17–19 September.
U.S. Department of Agriculture. 1997. Table III, Sources of Funds Allocated for Cooperative Extension Work, FY 1997, Washington, D.C.: U.S. Department of Agriculture.
U.S. Department of Agriculture. 1999. 1997 Census of Agriculture Vol.1, Part 51. Washington, D.C.: National Agricultural Statistics Service. U.S. Department of Agriculture.
U.S. Department of Agriculture. 1999a. Lincoln's Legacy, The Peoples' Department. Online. Internet. 5 August 1999. Available: http://www.usda.gov/yourusda/open.htm.
Wadsworth, Henry. 1999. Dr. Henry Wadsworth, then Director of Extension, Purdue University, Indiana. Personal conversation, June.
Ward, David. 1998. *University of Wisconsin-Madison, 1997–98 Annual Report.* Madison: Board of Regents of the University of Wisconsin.

Webb, Bud. 1997. South Carolina State Legislator and former Director of Cooperative Extension in South Carolina, at the 1997 National Public Policy Conference, Charleston, South Carolina, September.
Weiser, C.J. 1997. Broader Visions of Scholarship, Draft. Handout at workshop, "Scholarship Unbound—Reframing Faculty Evaluation and Rewards." Corvallis: Oregon State University, October 1998.
Weiser, C.J. 1999. Personal conversation, Corvallis, Oregon, 11 August.
Weiser, C. J. 1999a. Personal conversation, September.
Weiser, Conrad J. 1999b. Final Report to Kellogg Foundation, October 1998 OSU Workshop, Scholarship Unbound, Reframing faculty evaluation & rewards. Oregon State University.
Wilson, Charles. 1979. Reconciling National, International and Local Roles of Universities With the Essential Character of a University, in *Pressures and Priorities—Report of the Proceedings of the Twelfth Congress of the Universities of the Commonwealth*. London. The Association of Commonwealth Universities.
Wood, William W. 1978–79. Extensive personal conversations.

Index

n after page number refers to Notes.

Academic community
   misunderstanding of tenure by, 37–39
   public service and, 37–44
   scholarly reputation in, 40–44
Academic culture, need for change in, 180–89
Academic freedom, legitimate defense of, 37–38
Acupuncture, 41
AESOP, Inc., 119, 123
Agricultural constituents in society, changing power of, 56–57
Agricultural economics, 58–63
   finding cutting edge in, 60–61
Agricultural economists, responsibilities of, 61–63
Agricultural Experiment Stations, 99–100, 119
   research funded by, 22
Agricultural extension. *See also* Cooperative Extension Services
   agricultural economics and, 58–63
   evaluation of, in serving farmers, 94–96
   fit of technology transfer extension model in, 89–94
   need for economic and other social science input into, 196
   public advocacy role of, 104
Agricultural interest groups, 122
Agricultural interests, capture of cooperative extension by, 68–82
Agricultural research, linkage between U.S. Department of Agriculture and, in land-grant universities, 98–99
Agricultural Research Service (ARS), 114, 122

Agricultural sciences from theory to practice in, 48–64
Agriculture. *See also* American agriculture
   character of scientific advances being applied to, 86–87
   Cooperative Extension Service as hostage of, 72–81, 83–96, 196
   federal presence in, 87–88
   rural infrastructure serving, 86
Agriculture, U.S. Department of (USDA)
   Cooperative State Research, Education, and Extension Service in, 99, 113–15, 117, 120–22, 123
   education and, 8
   as federal partner in cooperative extension, 8, 97, 113–28, 197
   Food and Nutrition Service of, 126
   influence of, on Cooperative Extension, 98–99
   land-grant universities and, 69–70
   linkage between agricultural research in land-grant universities and, 98–99
   multiple bilateral relationships between, and the states, 99–100
   state dealing with, 118–20
Agriculture colleges
   administration of cooperative extension programs in, 20
   changed political economy of scholarship in, 53–63
   extension function and, 8–9
   funding changes in support of, 55–56
Alumni, public service program aimed at getting support from, 21–22
American Agricultural Economics Association, 59, 60

207

American agriculture. *See also* Agriculture
    forces of change in, 84–88
    growth in productivity of, 86
American Association of Agricultural
    Economics, 182
American Association of Land-Grant
    Colleges and State Universities, 128
American Farm Bureau, 51
American Farm Bureau Federation, 123
American Journal of Agricultural Economics,
    59, 61, 182–83
Analytic knowledge, 33
Association of County University
    Extension Councils, 189
Astrology, comparison of astronomy and,
    31
Astronomy, comparison of astrology and,
    31
Attribution condition, 194

Bailey, Liberty Hyde, 4
Bayh-Dole Act (1980), 175
Bok, Derek, 26–27

Cambridge University, 11, 16, 198
Carnegie Foundation, 17–18
    classification of universities by, 5–6,
        13–14*n*
    in evaluation of faculty, 46–47
Children, Youth, and Families at Risk, 117,
    118
Choices, 59, 182–83
Clarity in problem solving, 33
Classroom instruction, access to, 6–7
Clustering, 103–4
Collaborative-based research, 179
Commerce, U.S. Department of, extension
    services and, 124
    funds from, 126
Community-based research, 172, 179
Community-collaborative research, 172,
    173
Computer/information technology, 194
Computers, particularizing information
    on, 73
Consistency in problem solving, 32–33
Consumer risk, 85
Content transmission, 173
Cooperation extension faculty members,
    time allocation for, 49–50

Cooperative extension contacts,
    personalizing of, with farm audiences,
    73–74
Cooperative extension economists
    disciplinary contribution of, 63
    responsibility of, 61–63
Cooperative Extension Services, 3, 7, 24.
    See also Agricultural extension
    capture of, by farming interests, 71
    contribution to future of land-grant
        universities, 65–82
    control of agenda in, 76–77
    control of institutional setting of, within
        university, 77–81
    county support for, 97, 100–113
    deans of agriculture responsibility for
        in, 191–92
    emphasis on programs delivered through
        country extension offices, 25–26
    establishment of, at land-grant
        universities, 17
    federal support for, 97–98
    full time equivalent (FTE) staff
        employed by, 65
    as hostage of agriculture, 72–81, 83–96,
        196
    as information resource, 65–66
    in land-grant institutions, 20
    partnerships in, 97–130
    politics of, 129–30
    purposes of, 69, 113
    Smith-Lever Act in establishing, 52
    state support for, 97, 98–100
    traditional, 27*n*
    U.S. Department of Agriculture
        (USDA) as federal partner in, 8,
        98–99, 113–28
Cooperative extension specialists,
    contributions of, 63
Cooperative State Research, Education,
    and Extension Service (CSREES), 99,
    113–15, 117, 120–22, 123
    role in land-grant universities, 121–22
Cooperative State Research Service
    (CSRS), 113–14, 114
Core land-grant colleges, 48–49
    academic cultures in, 63–64
Cornell University, Cooperative Extension
    at, 171
Corporation for Public Broadcasting, 126

Corporation for Public University Outreach, 126–27
Country Life Commission (1910), 17
County Cooperative Extensions, 100–113
  campus faculty incentives, 105–6
  control of, in agenda, 101–2
  county and field staff context, 108–11
  county staff specialization and its management, 103–4
  county voices in, 111–13
  overcoming disconnect, 106–8
  programming disconnect, 102–3
  relationship between land-grant universities and, 108
  time frames for, 104–5
County Farm Bureaus, business activities of, 110–11

Dairy Herd Improvement Associations, 196
Deans of agriculture, responsibility of, in Cooperative Extension, 191–92
Deductive reasoning, 30
Development scholarship, 195
Dietary knowledge, impact on consumer demand, 85
Disintermediation, 84–85, 93
District of Columbia, University of, 3
Doctoral institutions, 6, 14*n*

Economics extension, 59–60
Education, separate and unequal philosophy of, 3
Education, U.S. Department of, and extension services, 124
Engaged land-grant universities, 169–90. *See also* Land-grant universities
  change in academic culture in, 180–89
  encouragement of consulting work at, 186
  faculty at, 185, 186, 187, 193
  guiding characteristics for, 189–90
  information technology at, 174
  institutional administrative structure of, 178–80
  involvement of whole university in, 174–77
  leadership of, 185–86
  manipulation of symbolism within, 193
  off campus involvement of facilities, 179–80
  people's desires from, 177–78
  students of, 170–74
Engagement, 169–90, 191–98. *See also* Engaged land-grant universities
  distinction between unidirectional model and, 26
  implied benefits of, 26
English model, 16, 17
Environmental grounds, scrutiny of farming on, 84
Environmental impacts of variety of farming practices, 93–94
Environmental Protection Agency (EPA), funds from, 126
Excess capacity benefit, 21, 25–26
Expanded Food and Nutrition Education Program (EFNEP), 171
  formula for, 99–100
Expanded Food and Nutrition for Low Income Families, 117–18
Experiment Station Committee on Policy (ESCOP), 100, 119
Extension Committee on Policy (ECOP), 100, 119
Extension economists
  disciplinary contributions of, 63
  responsibilities of, 61–63
External consistency, test of, 33

Faculty
  Carnegie Foundation in evaluation of, 46–47
  at engaged land-grant universities, 185, 186, 187, 193
  as no longer on cutting edge, 49–53
  scholarship in evaluation of, 45–46
  tenure protection for, 69
  time allocation of, 49–50
Falsification, 30, 33
Farm audiences, personalizing of extension contacts with, 73–74
Farm commodities, demand for, 84
Farm credit system, 68
Farmers, evaluation of agricultural extension programs in serving, 94–96
Farming
  environmental impacts of variety of practices, 93–94

Farming *(cont.)*
    scrutiny of, on environmental grounds, 84
Federal Agricultural Improvement Reform (FAIR) Act (1995), 84
Federal presence in agriculture, 87–88
Food stamp program, 126
Formula funds, decline in, to colleges of agriculture, 55–56
4-H programs, 67, 73, 74, 103, 109, 110, 117, 122, 131, 144, 156–61
    criticism of, 156
Freeman, Orville, 120
Full-time equivalent (FTE) extension staff in land-grant universities, 8–9
Funding of changes in agriculture colleges, 55–56

German universities, 5
    central concept of, 16, 17
GI Bill, 57
Glion Declaration, 178

Harvard University, 11, 28, 198
Hatch Act (1887), 7, 17, 52
Hatch funds, 190
Hawaii, University of, 146
Health and Human Service, U.S. Department of (HHS)
    extension services and, 124
    funds from, 126
Higher education
    access to, 4–5
    expanding demand for, 5–6
    monetary benefits of, 18
    nonmonetary benefits of, 18
    number of institutions, 5
    preservation of aristocracy and, 4
    vulgarization of, 5
Historical research, teaching to historical societies, 170
Historical societies, historical research teaching to, 170
History extension program, 169–90
House Agricultural Committee, 122
Housing and Urban Development, U.S. Department of, funds from, 126
Humanities, threat of demise of traditional, 175–76

Humanities extension at North Carolina State University, 132–38, 167, 170, 196
Humbolt University, 28, 198

Idaho, University of, 146
Illinois, county extension offices in, 111
Information technology at Engaged University, 174
Integrated pest management programs, 75–76
International markets, influence of, 85
Iowa, Cooperative Extension Services in, 112, 177
Iowa State University, 81, 146

Johns Hopkins University, 5
Journalists, Oregon extension communications staff as, 161–65

Kellogg Commission on the Future of State and Land-Grant Universities, 11, 36, 171, 172. *See also Returning to Our Roots: The Engaged Institution*
    authors of, 27
    establishment of, 27
    goals set forth by, 65
    leadership of, 12, 27
Kent State University, 146
Knapp, Seaman, 17
Kuhn, Thomas S., 29–31

Land-grant universities, 4
    See also Engaged land-grant universities
    academic cultures in core, 63–64
    at the beginning of the 21st century, 3
    changing size, program and governance of, 57
    classification of, 5–6
    commitments to instruction and service, 28–29
    contributions of Cooperative Extension to, 65–82
    control of county extension agenda by, 101–2
    control of the institutional setting of Cooperative Extension within, 77–81
    Cooperative Extension service programs, 7

core, 48–49
dilemma of, at end of 20th century,
    11–12
establishment of Cooperative Extension
    Service at, 17
farming people's support for, 9
full-time equivalent extension staff in,
    8–9
future of, 63–64
growth of, 57
imaging extension engaged, 169–90
as instruments of social change, 191
leadership of, 112
linkage between U.S. Department of
    Agriculture and agricultural
    research in, 98–99
new knowledge as main business of
    contemporary, 28–29
number of, 3
outreach functions of, 6–7, 9
as people's universities, 7–8
politics of, 197
problems in changing culture in, 193–94
public service function of, 28
public service programs of, 7, 15–27
    access to facilities, 22
    as by-product or spillovers, 22
    and excess capacity, 22–23
    necessary conditions for, 23–24
    research as, 22
    usefulness of, 25–26
    user fees in, 24–25
purposes of, 4, 51–52
relationship between county extension
    and, 108
role of Cooperative State Research,
    Education, and Extension Service
    (CSREES) in, 121–22
scholarly agenda in, 10–11
scholarship in, 5
service function of, 28
as source of baccalaureate and
    postbaccalauretae education, 5
supportive culture in, 194
tenure protection for faculty, 69
Liberal education, 16
Licensure of such products by university,
    195

Management, Analysis, and Planning
    Program (MAP), 74–75

Massachusetts, extension services in, 102
Massachusetts Institute of Technology, 11
Materials science center, funding of, at any
    university, 125
Mathematical models, 33
McGrath, C. Peter, 27
Michigan State University, county
    extension services at, 106–8
Minnesota
    extension services in, 103–4
    4-H program in, 157, 158
Minnesota, University of
    extension services at, 102
    Parents Forever extension program
        from, 152–55
    tenure issue at, 38
Missouri, University of, and St. Louis
    Storytelling Festival, 165–67
Monteclaire State University, 146
Morrill, Justin, 4
Morrill Land-Grant Act (1862), 3, 4, 7, 17,
    19, 56, 127
Morrill-Land Grant Act (1890), 3
Morrill-Wade Act (1862), 52

National Academy of Science, 193
National Association of State Universities
    and Land-Grand Colleges
    (NASULGC), 11, 12, 99–100, 119, 190
    Board on Human Sciences, 123
    Commission of Food, Environment, and
        Renewable Resources (CFERR),
        99–100
National Corn Growers Association, 121
National Education Association, on higher
    education, 28
National Farm Bureau Federation, 75–76,
    111
National 4-H Foundation, 127
National Institutes of Health (NIH),
    57–58, 125
National Research Initiative (NRI), 95, 121
    Competitive Grants Program, 91–92
National Science Foundation, 57–58, 121,
    125
National science policy, 57–58
New York County Extension Associations,
    111
New York state
    cooperative extension program in, 110

New York state *(cont.)*
  county extension educators in, 109
  4-H program in, 158
New York State College of Agriculture, 4
North Carolina State University, College of Humanities and Social Sciences, extension program in humanities and social sciences at, 132–38, 167, 170, 196
North Carolina Textbook Commission, 134
North Dakota, farm controls in, 88
Novartis Corporation, 175

Objectivity, tests of, 32–33
Ohio State University, 146
  leadership of, 12
Oregon State University, 81, 146
  Cooperative Extension at, 111, 180–81, 183, 193, 194, 197
  county extension services at, 106–8
  evaluation of scholarship on university-wide basis, 45–46
  extension communications staff as journalists, 161–65, 178
  leadership of, 12
  redefining scholarship and integrating extension field faculty at, 138–47
Outreach functions of land-grant universities, 6–7, 9
Oxford University, 11, 16, 28, 198

Parents Forever program, 152–55
Patenting of plant materials, 87
Peer review, 10, 41
Peer validation, 183
Pennsylvania State University, 77, 81
  Outreach and Cooperative Extension at, 192, 193
Plant Genome Project, 121, 130n
Plant materials, patenting of, 87
Pocket Pets, 156
Political capital, wasting of, 75–76
Politics of land-grant university, 197
Popper, Karl, 29–31
Pork industry and vertical integration, 93
Portland State University, 146
Portrait of Oregon, A, 164, 194
Pragmatism, 33

Proactive programming, 74–75, 142
Problems, distinction between practical and theoretical, 31
Problem solving
  clarity in, 33
  consistency in, 32–33
  research, 105
  values in, 32
  workability in, 33
Productivity, growth of, in American agriculture, 86
Product markets, market failures in, 56–57
Programming, reactive versus proactive, 74–75
Public advocacy role of agricultural extension program, 104
Public policy analysis and education, 179
Public service
  academic community and, 37–44
  bottom line for, 26–27
  contributions to scholars and scholarships, 29–37
  defined, 19–26
  direct, 20
  as enterprise category, 21
  function of the university, 28
  land-grant university interest in, 7, 15–27
  roots of traditional, 16–19
Publish or perish mythology, 40, 43, 62
Puritan ethic, 43
Puzzle solving, 30–31

Rationalism, 61
Reactive programming, 74–75
Relevancy, 34–37
Research
  collaborative-based, 179
  community-collaborative, 172, 179
  control of agenda, 53–63
  as function of the university, 28, 195
  funding of, by outside sources, 22
  pursuit of resources, 40–44
Researcher, distinction between scholar and, 34
Research universities, 5–6, 6, 13n, 29
  future contributions of, 26–27
  growth of, 9–10
  as source of knowledge, 10

Returning to Our Roots: The Engaged Institution, 11, 26, 63, 170–71, 189–90, 192–93, 197
Review of Agricultural Economics, 59
Rhode Island, University of, 98
Rockwell, Norman, 89
Roosevelt, Theodore, 17
Rural infrastructure, serving agriculture, 86

St. Louis Storytelling Festival, 165–67
 humanities extension about, 167
Schnittker, John, 120
Scholarly agenda, definition and refinement of, 10–11
Scholarly communities, 39–40
Scholars
 distinction between researcher and, 34
 public service and, 29–37
 reputation of, 40–44
Scholarship, 44–46
 changed political economy of, in colleges of agriculture, 53–63
 contributions of public service to, 29–37
 definition of, in evaluation of faculty, 45–46
 evaluation of, on university-wide basis, 45–46
 land-grant view of, 5
 political economy of, 40
Science
 growing distrust of, in Western societies, 86
 public service contributions to, 29–34
Science and Education Administration (SEA), 114
Scientific advances, character of, being applied to agriculture, 86–87
Scientific knowledge, development of, 30
Scottish universities, 16, 17
Separate and unequal philosophy of education, 3
Service function of the university, 28
Shapshot of Salmon in Oregon, A, 162
Smith-Lever Act (1914), 7, 17, 19–20, 20, 52, 89, 99, 110, 113, 190
 as amended through 1985, 113
Social contract
 establishing new, between university and community, 169
 renegotiating or abandoning, 191–98

Southern University System, leadership of, 12
Specialization, danger of, 34
Spillman, William J., 17
Spillover benefits, 21–22, 24
"Standards of Learning" in Virginia, 38–39
Stanford University, 11, 198
State Association of County University Extension Councils, 187, 188–89
State of the State, The, 185, 190
States
 dealing with U.S. Department of Agriculture (USDA), 118–20
 multiple bilateral relationships between U.S. Department of Agriculture (USDA) and, 99–100
 support for Cooperative Extension Services by, 97, 98–100
Student athletic programs, 21–22
Students of Engaged University, 170–74
Synthetic knowledge, 33

Technology transfer, 100–101, 179
Technology transfer extension model, 10
 fit of, in agricultural extension, 89–94
Tenure, 69
 academia's misunderstanding of, 37–39
Texas Agricultural Extension Service, 4-H program and, 157
Texas A & M, 146
Total quality management (TQM), 84, 93
Transformational learning, 171–72
Tuskegee University, 3

Unidirectional model, distinction between engagement and, 26
Universal product codes, use of, 85
University Extension Advisory Council, 189
User fees in public service programs, 24–25

Values in problem solving, 32
Verification, 30
Vertical integration and pork industry, 93
Veterans Administration, funds from, 126
Virginia Cooperative Extension Service, 74–75, 81
Virginia Farm Bureau, 81

Virginia Polytechnic Institute and State University, extension division at, 48–49, 58, 78, 81, 101
   cooperative extension at, 181
   Department of Agricultural Economics, 90–92
Virginia "Standards of Learning," 38–39
Virginia State University, 101
Vulgarization of higher education, 5

What We Heard and What We Did: Annual Report of Engaged University to the People of the State, 188
Wisconsin
   Community Resource Development (CRD), 147–52
   Cooperative Extension Services in, 112, 177
   County Extension Committee in, 109
   4-H program in, 156, 158–60
Wisconsin, University of, Extension, 196
"Wisconsin Idea," 17, 184
Wisconsin State Association of County Extension Committees, 112
Wissenschaft, 16
Workability, test of, 33–34, 36, 48
Written word, 39–40, 62
   need for extension purposes, 62–63
   reasons given for not using, 62, 142

Yale University, 11, 198
Youth and Families at Risk, funds from, 126
Youth At Risk, 122